THE POLITICS OF
THE OLYMPICS

THE POLITICS OF
THE OLYMPICS

A SURVEY

FIRST EDITION

Editors: Alan Bairner and Gyozo Molnar

Routledge
Taylor & Francis Group

LONDON AND NEW YORK

First Edition 2010
Routledge
2 Park Square, Milton Park, Abingdon, Oxfordshire OX14 4RN
711 Third Avenue, New York, NY 10017
First issued in paperback 2014

Routledge is an imprint of the Taylor & Francis Group, an informa business
© Routledge 2010

ISBN 978-1-85743-494-1 (hbk)
ISBN 978-1-85743-687-7 (pbk)

Development Editor: Cathy Hartley

Typeset in Times New Roman 10.5/12

Typeset by Taylor & Francis Books

Foreword

The editors of and contributors to *The Politics of the Olympics* share the belief that sport and politics are intimately connected and that nowhere is this relationship more apparent than in the Olympic Games.

This volume comprises a section of essays and an A–Z glossary. The collection of essays, which is preceded by an introduction written by the editors, sheds light on the discrimination, exploitation and conflicts that have been and continue to be associated with the Games. Chapters focus on topics such as the politics of the Olympic movement, the politics of hosting the Games, the politics of performance enhancement and the politics of gender discrimination. These issues are as old as the Games themselves. However, there are also more recent concerns, including the political links between disability and the Olympic movement, and between the Olympics and terrorism. Chapters on these particular themes are complemented by four studies that are specific to certain countries or regions—Germany during the rise to power of national socialism, Eastern Europe in the Cold War era, the Republic of Korea and Taiwan—with the overall aim of exposing and explaining the extensive links between politics and the Olympics.

To provide further information, an A–Z glossary, prepared by the editors, is included. This consists of around 90 entries, providing the reader with additional information about famous Olympians, presidents of the International Olympic Committee (IOC), specific events, and boycotts and demonstrations that have been of political significance in the history of the Games. The glossary does not include individual entries for the Paralympic Games, which are addressed fully in Chapter Five, nor the London 2012 Games. The latter are discussed at some length in the introduction, because most of what can be said at the time of writing is necessarily speculative, while the glossary is intended to provide factual information.

For reference, a map is included that shows the venues of the Olympic Games, emphasizing the extent to which they have remained very much a Western phenomenon, in spite of celebrated exceptions to that general rule. To assist in further research into the often complex topics discussed, chapter authors have provided bibliographies and further reading lists at the end of chapters.

Ultimately, the main aim of the book is to look beyond the façade of the IOC and the Olympics themselves and to debunk some of the myths and traditions that surround them.

Alan Bainer and Gyozo Molnar
January 2010

Contents

List of tables

Acknowledgements

The editors wish to express their thanks to the contributors for agreeing to write chapters, for the quality of their work and for their capacity to respond quickly and with good humour to our incessant demands. We are also extremely grateful to Cathy Hartley at Routledge for her initial approach, her unwavering support and her much needed guidance. The editors are also grateful for the technical support provided by Paola Celli and Alison Neale during the production process.

Thanks are also due to the Cartographic Unit, University of Southampton, for providing the map of Olympic venues.

The editors and contributors

Alan Bairner is Professor of Sport and Social Theory at Loughborough University (UK), having previously worked at the University of Ulster for 25 years. He is co-author (with John Sugden) of *Sport, Sectarianism and Society in a Divided Ireland* (1993) and author of *Sport, Nationalism and Globalization: European and North American Perspectives* (2001). He edited *Sport and the Irish: Histories, Identities, Issues* (2005) and was joint editor (with John Sugden) of *Sport in Divided Societies* (1999) and (with Jonathan Magee and Alan Tomlinson) of *The Bountiful Game? Football Identities and Finances* (2005). He serves on the editorial boards of the *International Review for the Sociology of Sport* and the *International Journal of Sport Policy*.

Gyozo Molnar is Senior Lecturer in Sport Studies in the Institute of Sport and Exercise Science at the University of Worcester (UK). He completed his doctorate in the sociology of sport at Loughborough University, was the co-ordinator for the Centre for Olympic Studies and Research, and has taught modules in the areas of sociology and politics of sports and exercise. His current publications and research revolve around migration, football, globalization, national identity, the Olympics and sport-related role exit.

David L. Andrews is Professor of Physical Cultural Studies in the Department of Kinesiology at the University of Maryland (USA) at College Park and affiliate faculty member of the Departments of American Studies and Sociology. He is assistant editor of the *Journal of Sport and Social Issues*, and an editorial board member of the *Sociology of Sport Journal, Leisure Studies*, and *Quest*. His research utilizes various theories and methods drawn from sociology and cultural studies to critically analyse the relationship between broader social structures and the embodied representations and experience of contemporary physical culture (including sport-, exercise-, health- and movement-related practices).

Susan J. Bandy is currently a visiting professor in the Department of Physical Activity and Education at the Ohio State University (USA). Her research interests include sport literature, the history of women in sport, and gender and the body in sport, including issues related to transsexuality, transgender and intersexuality in sport. She has published several books, including: *Coroebus Triumphs: The Alliance of Sport and the Arts* (1988), *Crossing Boundaries: An International Anthology of Women's Experiences in Sport*

(1999), and *Gender, Body and Sport in Historical and Transnational Perspectives* (2007). She is currently working on a book devoted to Scandinavian sport literature entitled *Fortællinger om idræt i Norden: Helte, erindringer og identitet (Scandinavian Sport Literature: Heroism, Memory and Identity)*.

Rob Beamish has held a joint appointment in the Department of Sociology and the School of Kinesiology and Health at Queen's University, Kingston (Canada) for the past 25 years. In addition to *Fastest, Highest, Strongest: The Critique of High-Performance Sport* (2006), co-authored with Ian Richie of Brock University, and a forthcoming introduction to sociology text, *Sociology's Task and Promise*, he has published numerous articles and book chapters on classical sociological theory and on high-performance sport.

Miklós Hadas was awarded a PhD in sociology from the Hungarian Academy of Sciences and is Professor of Sociology and Co-Director of the Centre for Gender and Culture at Corvinus University, Budapest (Hungary). From 1990–2002 he was the founding editor-in-chief of *Replika*, a leading Hungarian journal of social sciences. He is author of numerous articles on gender and masculinity and of *A modern férfi születése* (The Birth of Modern Man) (2003). This book was awarded the Polányi Prize by the Hungarian Sociological Association for the best sociological book to be published that year.

John Horne is Professor of Sport and Sociology in the School of Sport, Tourism and the Outdoors, at the University of Central Lancashire (UK). He is managing editor of the journal *Leisure Studies* and a member of the editorial boards of the *International Review for the Sociology of Sport* and *Sport in Society*. His publications include numerous articles in peer-reviewed journals and book chapters, and, as author, *Sport in Consumer Culture* (2006), as co-author, *Understanding Sport* (1999), and as co-editor, *Sports Mega-Events* (2006), *Football Goes East: Business, Culture and the People's Game in China, Japan and Korea* (2004), *Japan, Korea and the 2002 World Cup* (2002) and *Sport, Leisure and Social Relations* (1987).

P. David Howe is Senior Lecturer in the Anthropology of Sport in the School of Sport, Exercise and Health Sciences at Loughborough University and Deputy Director of the Peter Harrison Centre for Disability Sport. Trained as a medical anthropologist, he is the author of *Sport, Professionalism and Pain: Ethnographies of Injury and Risk* (2004) and *The Cultural Politics of the Paralympic Movement: Through the Anthropological Lens* (2008).

Jung Woo Lee is currently a part-time lecturer and researcher at Sungkyunkwan University, Republic of Korea (South Korea), and holds a PhD in the sociology of sport from Loughborough University. His research interests include media sport, North and South Korean relations, communism, nationalism and globalization. Recently, he has published on South Korean sporting nationalism and on the political nature of North Korean sports in the *International Review for the Sociology of Sport* and the *Journal of Sport and Social Issues*, respectively.

Ping-Chao Lee is Associate Professor at the Department of Physical Education, National Taichung University in Taiwan. He received his PhD from the Institute of Sport and Leisure Policy, Loughborough University and has published papers both in English and Chinese on the challenges facing the Taiwanese sporting system. His current research interests include the political aspects of the Olympics and the field of governance of professional baseball in Asia.

Helen Jefferson Lenskyj is Professor Emerita at the University of Toronto (Canada). Of her nine books, three present critiques of the Olympic industry: *Inside the Olympic Industry: Power, Politics, and Activism* (2000), *The Best Olympics Ever? Social Impacts of Sydney 2000* (2002) and *Olympic Industry Resistance: Challenging Olympic Power and Propaganda* (2007). She is currently co-editing a critical anthology of Olympic studies with Stephen Wagg, as well as continuing her work as an anti-Olympic activist in Canada.

Jaime Schultz was awarded a PhD from the University of Iowa in 2005. She is an assistant professor in the University of Maryland's Physical Cultural Studies programme. The author of many chapters and journal articles, she is currently completing two books: *Moments of Impact*, on race, cultural memory and sport history, and *From Sex Testing to Sports Bras: Gender, Technology, and US Women's Sport* for the University of Illinois Press.

Michael L. Silk is a Senior Lecturer in the Faculty of Humanities and Social Science at the University of Bath (UK). His research and scholarship centre on the production and consumption of space, the governance of (physical) bodies and the performative politics of identity within the context of neo-liberalism. His work on identity politics, the Olympic Games, and other mega-events has recently been published in *Media, Culture, Society*, the *International Journal of Media & Cultural Politics*, *Studies in Ethnicity & Nationalism*, the *Sociology of Sport Journal*, and *Sport in Society*.

Tien-Chin Tan is assistant professor in the sociology of sport and policy at the National Taiwan Normal University. He completed his PhD in sport policy at Loughborough University in 2008. His main research interests are public policy for sport, particularly in the areas of sport development, elite sport development; youth sport; and school-based sport policy in Taiwan and the People's Republic of China. He has published articles in various journals, including the *International Journal of the History of Sport*, the *International Journal of Sport Policy*, the *Journal of Sport and Social Issues* and *The China Quarterly*.

Christopher Young is Reader in Modern and Medieval German Studies, and Head of the Department of German and Dutch at the University of Cambridge (UK) and Fellow of Pembroke College. He has authored and co-edited eight books on German language, literature and culture, and a further six volumes and journal special issues on international sport, including (with Alan Tomlinson) *German Football* (2006) and *National Identity and Global Sports Events* (2006). In 2010, *The Munich Olympics 1972* and *The Making of Modern Germany* (with Kay Schiller), for which he received support from the AHRC (Arts and Humanities Research Council), British Academy and the Alexander von Humboldt Foundation, will be published by the University of California Press. Most recently, he co-edited a special issue of *German History* (2009) on the history of German sport.

Abbreviations

a.m.	ante meridiem (before noon)
BAU	Business as Usual
BBC	British Broadcasting Corporation
CA	California
CD ROM	Compact disc read-only memory
CEO	Chief Executive Officer
CO	Colorado
CT	Connecticut
DC	District of Columbia
DNA	Deoxyribonucleic acid
DSB	Dispute Settlement Body
EC	European Commission
ECOSOC	United Nations Economic and Social Council
Est.	Estimate/estimated
EU	European Union
FAO	Food and Agriculture Organization
FBI	Federal Bureau of Investigation
GATT	General Agreement on Tariffs and Trade
GDP	Gross Domestic Product
GDR	German Democratic Republic
ha	hectare
IBRD	International Bank for Reconstruction and Development (World Bank)
IF(s)	International sporting federation(s)
IL	Illinois
IMF	International Monetary Fund
IOC	International Olympic Committee
kg	Kilogramme
KS	Kansas
MA	Massachusetts
MD	Maryland
MDGs	Millennium Development Goals
MNC	multinational companies
NGO	non-governmental organization
NJ	New Jersey
NOC	National Olympic Committee

OCOG	Organizing Committee(s) for the Olympic Games
OECD	Organisation for Economic Co-operation and Development
PA	Pennsylvania
POW	prisoner of war
PR	public relations
TEU	Treaty on European Union
TX	Texas
UK	United Kingdom
UN	United Nations
UNESCO	United Nations Educational, Scientific and Cultural Organization
UNGA	United Nations General Assembly
US(A)	United States (of America)
USSR	Union of Soviet Socialist Republics
vs.	versus
WADA	World Anti-Doping Association
WHO	World Health Organization
WTO	World Trade Organization

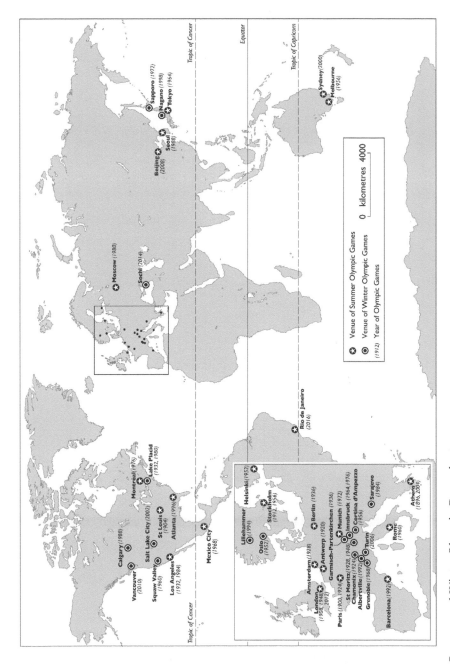

Map Summer and Winter Olympic Games host venues

Essays

Essays

The politics of the Olympics: An introduction

ALAN BAIRNER AND GYOZO MOLNAR

INTRODUCTION

When Rio de Janeiro was awarded the right to host the 2016 Summer Olympics in the face of apparently strong competition from a Chicago bid that had been enthusiastically endorsed by US President Barack Obama, Dick Pound, a Canadian International Olympic Committee (IOC) member commented, 'It is not an anti-America thing or an anti-Obama thing. It's a sports competition, not true politics' (*Sunday Morning Post*, 4 October 2009). In Pound's words were echoes of sentiments expressed in 1936 by an illustrious predecessor in the Olympic Movement. Faced with demands that there should be a boycott of the Olympic Games in Berlin, Avery Brundage, the President of the US Olympic Committee and later the President of the IOC, protested that 'politics has no place in sport'. It is a mantra that has been repeated over the years by countless sports leaders, although increasingly such a claim is likely to be met by muted laughter at the very least. Indeed, the IOC's Olympic Charter stresses the need 'to oppose any political or commercial abuse of sport and athletes'. However, as is clearly demonstrated in each of the chapters in this book, sport and politics do mix and it is inevitable that they should do so. Indeed, this is the case regardless of how we define the political.

There are various ways in which what is political can be defined. It is reasonable, for example, to argue that every human association is inherently political since each incorporates processes through which conflict is managed, resources allocated, rewards conferred and punishments prescribed and enacted. Sports clubs and organizations are no different from other human associations in this respect, despite what we have been often led to believe. More often than not, however, the study of politics is centred on the activities of states as opposed to other social organisms for the simple reason that the state is the most common, the most genuinely universal, human association (Berki, 1977). Furthermore, it demands our highest loyalty and usually receives it. It has a stricter and more pronounced differentiation of roles and functions than most other associations. Finally, amongst all human associations, the state presents us with the greatest intellectual problems. Inevitably, therefore, any true political reading of sport must take into account its relationship with states at both domestic and international levels. What Antonio Gramsci (1971), more than most political thinkers, showed us is the degree to which politics intrudes on every facet of our lives. Thus, we can move relatively easily from the politics of human associations through the politics of the state to the politics of just about everything, and all of these definitions of the political have relevance for the study of the Olympics.

In 2002, at the Salt Lake City Winter Games, Alain Baxter became the first British competitor to win a medal (bronze) in alpine skiing. This achievement was subsequently overshadowed when Baxter was found guilty of taking a performance-enhancing drug. He successfully appealed this decision but did not win the right to have his medal returned. These events not only overshadowed Baxter's performance on the slopes but also another controversial facet of his Olympic appearance. Prior to competition, he had dyed his hair blue and white in the shape of the Cross of St Andrew, the national flag of his native Scotland. In accordance with IOC rules, the British Olympic Committee asked Baxter to remove the dye which could be deemed to be symbolically political—in the sense that Baxter was making a statement about Scotland's non-independent, constitutional status as a constituent part of the United Kingdom of Great Britain and Northern Ireland on behalf of which he was competing at the Games. It is almost certain that the gesture owed far more to unthinking patriotism than to a carefully orchestrated campaign to promote the cause of Scottish nationalism, but it was sufficient to be regarded as political by the Olympic authorities. When real political statements are made by athletes, the reaction, not surprisingly, can be even more punitive and nowhere has this been more apparent than in the context of the politics of 'race'.

THE OLYMPIC GAMES AND THE POLITICS OF 'RACE'

The Olympic Movement has been most directly associated with and affected by 'race' issues in three main ways. First, there have been debates in the IOC about institutionalized racism within member countries. In the case of Nazi Germany, a proposed boycott of the Berlin Games was avoided despite widespread knowledge of the ruling regime's racist policies which extended into the world of sport itself. Similarly, the Beijing Games of 2008 went ahead regardless of concerns about the Chinese Government's treatment of ethnic minorities. In the case of South Africa, on the other hand, after the passing of a United Nations General Assembly Resolution in 1962, the IOC agreed to take part in an international boycott in protest against that country's racist apartheid policy. Having competed in the Olympics for the first time in 1904 and having participated in the Rome Games in 1960, South Africa was prevented from taking part in further Olympics until 1992, by which time the apartheid system was in the process of being dismantled.

Second, the Olympic Games have been used as a stage upon which to protest directly against racism. At the 1968 Games, African-American Jim Hines, who won the 100 metres in world record time, refused to accept his medal from Avery Brundage, the leading figure in his country's Olympic Movement and the man who had spoken out against a boycott of the Berlin Games (Witherspoon, 2008). Then, on 16 October, during the medal ceremony for the men's 200 metres, two American medal winners, Tommie Smith (gold) and John Carlos (bronze), raised black-gloved fists whilst the American national anthem was being played—'the most indelible image from the 1968 Mexico City Olympic Games' (Witherspoon, 2008: ix). Accompanying them on the winners' podium during their 'Black Power' salute was an Australian athlete, Peter Norman, who demonstrated his support for Smith and Carlos by wearing the Olympic Project for Human Rights badge. Smith and Carlos received lifetime IOC bans for using the Games to make a political point. Norman was omitted from the Australian Olympic team in 1972. The actions of Smith and Carlos were aimed at ongoing racial discrimination in the USA and in support of the campaign for Civil Rights for Americans of colour.

Much earlier, the African-American hero of the Berlin Games, Jesse Owens, is said to have claimed that it was not Hitler's response to his achievements that upset him but that of his own President, and, much later, Cassius Clay was clearly unhappy about returning to the USA as an Olympic hero only to be treated as a second-class citizen. Kathy Freeman's performance at the Sydney Games and her subsequent celebration which involved both an Aborigine flag and an Australian one might be interpreted as a less direct, more subtle, comment on racist practices in her home country.

Finally, discussions within the academic community about Olympic performances have also raised the spectre of 'race', with some scholars arguing the case for a modified form of biological determinism with reference to the success of sprinters of west African origin and distance runners from north and east Africa. Whilst repudiating the charge of biological determinism, Jon Entine (1999) has sought to ensure that the significance of physiological difference is not ignored when assessing elite sport performance. However, such arguments have been refuted (on occasion even described as racist) by those who fear their wider implications and argue instead that socialization, hard work, role models and access are more important explanatory factors. According to John Hoberman (1997: 99), 'the ascendancy of the black athlete over the past century and the growing Western belief in his biological superiority represented a historic reversal of roles in the racial encounter between Africans and the West'. In fact, the reversal has only been partial in that, whilst there may have been a willingness to concede that the white male possesses no innate physical superiority, athletic prowess has been used to endorse the myth of the intellectual inferiority of the black male—a myth that has been perpetuated in no small measure by what Hoberman accurately refers to as 'the African-American Sports Fixation' (p. 3). Here, then, are examples of political debate about 'race' unfolding within the context of the Olympics. It can also be argued, however, that the Olympic Movement and the mega-events that it promotes are themselves inherently political.

THE POLITICS OF THE BEIJING OLYMPICS

The Olympic Movement has come a long way since its foundation in 1894. Its initial features, goals and values have been altered and remoulded, sometimes slightly but occasionally more drastically, to cater for the needs of the powers that be. The number of participating nations, athletes and sporting events involved in the Games has shown an extraordinary growth. The media attention given to the Games and the financial incentives of hosting and advertising in and with the Games have reached astronomic proportions. The global presence and significance of the modern Olympic Games is ever increasing such that they have now become the world's largest and most popular quadrennially recurring sporting event. In the face of all these changes, there is one aspect of the Games that remained steadfast—the political connections the Olympics have to wider national, regional and global socio-political issues and tensions. In other words, 'the Olympic movement was never free from the effects of economic and political power … [and] no nation will pass over the opportunity to use the Games to make a political statement' (Tomlinson and Whannel, 1984: v–vii).

The most recent example of politics interfering with the Games, or vice versa, was provided by the 2008 Summer Olympics in Beijing. One would seriously struggle to find an aspect of these Olympics which did not have some connection with regional or global politics (Jarvie *et al.*, 2008). China began toying with the idea of bidding for and, if successful, hosting the Games as early as the 1940s. However, unfortunate political events (including civil war) and resultant financial limitations prevented an actual bid from being submitted. Thereafter political tensions between Taiwan and the People's Republic of China led initially to the latter being excluded from the Olympic family. After the separation of Taiwan and China in 1949, China's relationship with the IOC became erratic, with China boycotting a number of Games. In fact, it was only after Mao Zedong's death that China could fully reengage with the modern world (Guoqi, 2008). Subsequently, in the 1980s, plans to bid for the right to host the Olympics in China began to be drawn up. Guoqi explains that this was the consequence of 'an obsession with attaining international prestige, expressing Chinese national pride, and overcoming historical frustration' (Guoqi, 2008: 233). In 1992, Beijing submitted a bid to host the 2000 Summer Games. Although China was in the lead during the IOC bidding procedure, in the fourth round Sydney won the right, on 23 September 1993, to stage the 2000 Games. Because of the unexpected outcome, the Chinese Government and the Chinese Olympic Committee suspected political machinations and were displeased with the final IOC decision. Chinese urban youth and intellectuals shared this view and believed that a Western plot, led by the USA, had prevented China from securing the right to stage the Olympics (Guoqi, 2008). As a consequence of this defeat, Beijing did not bid for the 2004 Games, but went for the 2008 Olympics instead, securing hosting rights by a large margin.

After the initial excitement of winning the right to stage the Games, the nature and magnitude of the task ahead began to materialize and China became more and more exposed to the watchful eye of the Western media and their audience. Stories were reaching the global audience about the unfair treatment of those people who were forced to vacate their houses for the sake of building Olympic venues in Beijing. Moreover, the Beijing Games became the focus for world-wide protest involving demonstrators concerned with China's treatment of Tibet and with human rights violations within the country as a whole. In addition, the fact that the Taiwanese authorities refused to allow the torch to be carried through Taiwan because there was a feeling that the impression that the Chinese were seeking to create was that this would be part of the domestic leg of the torch's journey rather than being clearly international, did not help the situation.

The IOC hardly ever concerns itself with the sufferings of the minority or of disadvantaged populations. Its members cling instead to the principles that they can showcase most effectively in the wonderfully choreographed spectacle of the Olympic Games, where the glamour and glitter of athletic achievements and perfectly sculpted athletic bodies far outweigh the personal troubles of those who have been exploited, abused and dehumanized for the sake of the largest sporting enterprise. In this sense, despite the political difficulties that had preceded them, the Games in Beijing were indeed the 'best Olympics ever'.

TOWARDS LONDON 2012

In 2012, London will become the first city to host the Olympic Games for a third time. The circumstances in which these Games will take place are painfully ironic. In both 1908 and 1948, London came to the rescue of the Olympic Movement. From relatively inauspicious beginnings, the London 1908 Games have come to be widely recognized as the first modern Olympics in any meaningful sense. The Games were initially scheduled to take place in Rome. However, the eruption of Mount Vesuvius in 1906 meant that funds had to be diverted to disaster relief for the city of Naples. London came to the rescue with only 18 months to prepare—a remarkably short time when one considers how long host nations now have to ready themselves. Nevertheless, the White City Stadium was hastily constructed and the Games went ahead on time thanks in no small measure to the organizational abilities of Baron Desborough of Taplow (Kent, 2008).

The London Games of 1948 have become known as the 'austerity Olympics' because of the economically challenging post-Second World War conditions in which they took place (Hampton, 2008). The United Kingdom was still in a state of slow recovery. Bomb sites littered the urban landscape. Food and other staple products, from eggs to petrol, were strictly rationed. In such circumstances, it is scarcely surprising that controversies dogged the organization of the 1948 Games.

Because of the grim socio-economic conditions of the time, the press chiefly adopted a defeatist attitude and questioned both the relevance of the modern Olympics and the likely success of London as a host city. It appeared that the press critique was not entirely speculative. The organization of the Games was under-funded and progress in building a satisfactory infrastructure was slow.

In addition to domestic organizational difficulties, the London organizing committee was facing other problems as well. Invitations to potential participating countries and their replies were often delayed due to the ineffective international mail service. Furthermore, like the United Kingdom, many of the potential participant countries were still adversely affected by the hardships of post-war conditions and, thus, were slow and inefficient in letting the organizers know key details such as arrival times and numbers of athletes. Therefore, the organizers had to deal with a plethora of late and inappropriate entries and schedule events without knowing the exact number of athletes that would be taking part in them. In the end, in spite of the difficulties, the 1948 Games were a relative success. Once again, it seems, London had saved the day.

From the outset the outlook for the 2012 Games seemed so much more propitious. Unlike in 1908, there would be no need for hasty preparations. Furthermore, there would be none of the austerity that had beset the 1948 Games. 2012 was going to be so different, delivering urban regeneration and a sporting legacy in a confident, multicultural and prosperous United Kingdom at ease with itself thanks to its New Labour leadership. It is ironic, therefore, that in at least two important respects, the 2012 Games may yet turn out to be the Olympics (and Paralympics) that London could not afford to host.

When, in 2005, the city won the right to host the Games, the only people who were talking seriously about the collapse of the world banking system and of global recession were a dwindling band of academic Marxists. In the United Kingdom, sterling was strong, house prices were so high that many saw property as a major source of personal financial security, and unemployment was low. Tony Blair, charismatic and vacuous in almost equal proportions, was presiding over 'Cool Britannia'.

By the end of 2008, however, with Gordon Brown having replaced Blair as Prime Minister, the world was an altogether different place. Banks and building societies had collapsed or had only been saved from collapse by a massive injection of public funds. The value of sterling had plummeted in the international money markets. Property prices were falling as unemployment rose and fears about pensions increased. Public spending on other projects would continue to be under threat regardless of which political party would be in power by 2012. For those who dislike sport, and also for many lovers of sport, spending public money to host the Olympics in such circumstances was folly at best and evil at worst. Could the United Kingdom afford the Olympics? The economic challenge is undeniable as is the other major threat to the Games.

Whereas Barack Obama's attempt to bring the Olympics to Chicago failed, Tony Blair played an important part in securing the Games for London. His persuasive rhetoric undeniably helped London to win the votes needed to see off the challenge of Paris. Yet, it can also be argued that Blair's foreign policy, particularly his support for the USA in Iraq and Afghanistan, has made London a potentially very dangerous place to hold the Olympics.

On 7 July 2005, four bombs exploded on the London transport system, killing 52 commuters and four suicide bombers with probable links to Al-Qa'ida. The bombings took place the day after the IOC had announced that London had won the right to host the 2012 Games. The timing may well have been coincidental. The bombings demonstrated, however, that security would be a major issue for the London 2012 organizers. Indeed, speaking in 2006, the then Metropolitan Police Commissioner, Ian Blair, commented that there could be no doubt that the 2012 Games, if the current threat scenario remains the same, would be a huge target and that people would have to understand that and work on that basis. With British troops still engaged in an increasingly bloody conflict in Afghanistan as we write, it can legitimately be suggested that the 'current threat scenario' does remain. How the politics of the so-called 'war on terror' might impact on the 2012 Olympic Games remains to be seen.

Although security and economic factors are the main considerations for the local organizers, it should be noted that questions have also been asked concerning the extent to which the United Kingdom beyond London and the south-east of England will benefit from the Games. It is important to remember that when the Olympics were previously held in London—in 1908 and 1948—the United Kingdom was a considerably more united political entity than it is today. In the period that elapsed between the first and second London Olympics, the Irish question was partially solved through partition. Otherwise the United Kingdom continued to be administered as a collective unit from London. At present, however, Northern Ireland, Scotland and Wales each has a devolved administration and greater power has been granted to the English regions and, indeed, to London itself and its elected Mayor. Discussions as to whether the United Kingdom would be able to field an Olympic football team in 2012 were undeniably heavily influenced by football-related issues and, in particular, by fears on the part of the sport's governing bodies in Northern Ireland, Scotland and Wales that their current independence could be threatened. It is important to recognize, however, that this debate (and the resultant decision that the United Kingdom should be represented by an England-only team) has also taken place within a political context in which the United Kingdom is arguably less united than at any point since the mid-18th century. In such circumstances, it will be difficult to ensure that London's Games will be widely recognized as the United Kingdom's Games.

Finally, there is the question of legacy. In terms of urban regeneration, the task of transforming a rundown area of east London seems daunting if not impossible (Gibson, 2009a). In addition, there are growing concerns about the sporting legacy that can be accrued from the Games, with participation levels falling rather than rising and fears that after the Olympics, regardless of how British competitors perform, investment in sport will be drastically cut

(Gibson, 2009b; Gibson, 2009c). In sum, as we write, it is difficult to envisage that London 2012 can become the 'best Olympics ever'.

THIS BOOK

In addition to being informative, the main aim of this collection of essays and glossary is to engage in critical dialogue about the Olympics by raising awareness and providing a politics-focused account of the Movement and its sponsored events that we are all expected to love. The objective is not to promote antagonism towards the concept and events of Olympism, but nor is it to 'toe the celebratory party line'. We do not necessarily deny the potential benefits of taking part in and hosting the Olympics Games, but, along with the contributors to this volume, we suggest that to believe in the 'magical social healing power' of the Olympics is not only naive but demonstrates a lack of basic knowledge of the history of the modern Olympics. In this volume, we aim to look beyond the façade of the Olympics and debunk some of the myths of kudos, glory and tradition. Consequently, the chapters in this anthology illustrate the strong but changing links between the modern Olympic Games and politics, in general, and address and discuss the key political aspects and issues in relation to the Games themselves, to national and international sport organizations, and to specific countries' attitudes towards (ab)using the idea/ ideal of the Olympics for their own political agenda.

The first chapter provides a general but thorough overview of the internal politics of the Olympic Games and the Olympic Movement. Helen Lenskyj argues that the Olympics are 'one of the most powerful as well as one of the most under-scrutinized business enterprises in the world' and, consequently, uses the term 'Olympic industry' to characterize the nature of this particular sporting arena. Lenskyj sheds light on the internal intricacies of the IOC and gives a critical, political insight into the largest recurring sporting event of modern times.

The next piece, by John Horne, addresses a particular political aspect of the modern Olympics, i.e. hosting the Games. Initially, it may appear to some that playing host to the Olympics has great benefits for the hosting nation/city, including financial, social and political gains. However, as Horne suggests, staging the Games is much more complex and there are both costs and benefits to consider. In explicating the socio-political complexities involved in deciding to host, in bidding, in winning the right to host and in hosting, Horne pro-vides analytical answers to questions such as: Why do governments and cities compete for the right to host major international sporting events such as the Olympics?; what are the commercial underpinnings of hosting the Olympic Games?; and how do Olympic 'boosters' and 'sceptics' respectively portray the 'legacies', economic and otherwise, that are proclaimed for the Games?

Another political issue that surrounds the Olympics and the IOC, in parti-cular, is related to the politics of gender, more specifically women's (under) representation in the Olympics and Olympic Movement both as athletes and as IOC/NOC (National Olympic Committee) representatives. Considering that no women were allowed to compete at the first modern Olympic Games in 1896 and that sport itself was and remains renowned for being a male preserve, the feminist movement in relation to female involvement in organized elite sports, has achieved significant success. Nevertheless, while acknowledging this development in the third chapter, Susan Bandy observes that there 'are still gaps in the participation rates of males and females ...', and countries continue to send teams to the Games without female representation'. For that reason, Bandy focuses on the changing nature of women's involvement in the Olympics thereby offering a historical account of women's oppression and marginalization in the modern Olympics.

Women's role and involvement in the Olympics is only one of many pressing and controversial political issues associated with the modern Games. One of these is the use of illegal performance-enhancing substances in elite sports. Rob Beamish in the next chapter considers the politics of performance enhancement in the Olympics. The issue itself is so contradictory that it has often divided IOC members on the fundamental questions of what is performance enhancement and what is sport. In this chapter, the key ethical and political controversies and concepts are addressed and discussed in a historical framework underpinned by the notion of modernity.

The age of modernity has not only put our views on drug use in sport to the test, but it has also challenged the very idea and raison d'être of sport, including sport for disabled people. Even though mentally and physically challenged athletes have been competing in international sport for a long time, the general public is still often led to believe that sport events for disabled people are devoid of politics and above petty political argument. As David Howe explains with specific reference to the relationship between disability, Olympics and Paralympism, 'This myth of goodness associated with sport for the disabled should be critically examined'. In Chapter Five, Howe explores a number of key developments within disability and Paralympic sports that have influenced the politics of the Paralympic Movement, paying particular attention to the issue of governance.

As all the chapters illustrate, politics are integral to the Olympics, whether this is a case of the Games influencing external political developments or of political incidents having an impact on the Games. One particular way in which the Olympics have been affected by politics is through terrorism, which poses an ever-growing threat to contemporary society. Although the Games had been threatened with potential terrorist attacks even earlier, it was in 1972 in Munich that the Games were first interrupted by terrorism, as balaclava-wearing Black September terrorists kidnapped and subsequently killed members of the Israeli Olympics team. In the sixth chapter, having commented on the events of the 1972 Munich massacre, David L. Andrews, Jaime Shultz and Michael L. Silk explain that the relationship between terrorism and the Olympic Games is more diverse and complex than one might initially think. Consequently, their aim is 'to complicate popular understandings of

11

terrorism, through recourse to its various, and varied, Olympic manifestations'. To this end, they discuss the importance of recognizing the difference between violent acts that are directed at the state (or host nation) and those that are perpetrated by the state (or host nation).

Violence perpetrated by the state against Olympic practices and ideals is further investigated by Christopher Young through a detailed case study of the 1936 Berlin Games. Young admits that much has already been written about these Games, especially about Hitler's use of the event for propaganda purposes and about the role of Jesse Owens who almost single-handedly undermined the Nazi ideology of Aryan physical supremacy. He argues, however, that most previous accounts are reductive. In an attempt to debunk some Olympic myths, Young provides a comprehensive socio-historical account of the Berlin Games that contextualizes the event both historically and politically, and sheds a different light on the Games, on the links between Germany and the IOC, and on the racist attitudes and practices of the host nation.

In Chapter Eight, Miklós Hadas develops a Central and Eastern Europe-focused account of the relationship between the Cold War and the Olympics. Hadas acknowledges the distinctiveness of the Games in providing 'a unique opportunity to examine the interplay of complex symbolic forces on the global scene'. After briefly reviewing the key aspects of the Cold War, its effects on the Olympics and the symbolic fight between political 'good and evil' in the international sporting arena, Hadas turns his attention to the real and symbolic investments the countries of the communist bloc made in order to ensure medals and athletic domination for their sportsmen and sportswomen. Although from the outside, the USSR communist bloc looked healthy and largely coherent, there were various intra-bloc tensions, one of which manifested itself in the excessive aggression and bloodshed witnessed during the 1956 Melbourne Games and the now infamous water polo match between the USSR and Hungary. An account of this clash of nationalistic sentiments is situated within the context of rebellion and resistance to the aggressive Sovietizing practices within the Central and Eastern European region.

To further investigate the Cold War's impact on politics and sport, the two states on the Korean peninsula and their deep-seated political tensions are discussed next. Jung Woo Lee explores the ever-changing political relations between the Democratic People's Republic of Korea (North Korea) and the Republic of Korea (South Korea) and how these are manifested in the sporting arena, from the 1988 Seoul Games to the Beijing Olympics of 2008. Lee argues that, even though the Olympics have always had political implications for the two Koreas, the latter 'have barely used sport as a vehicle for touting the supremacy of their respective political ideologies'. On the other hand, the Olympic Games have provided a context for the representation of unified or separate national identities, depending on the fluctuating political undercurrent on the Korean peninsula.

The final chapter focuses on the relationship, or in this case, the direct impact, of the Beijing Games on another east Asian country—Taiwan or, officially, the Republic of China. Because the event itself took place relatively recently and the data presented here have been analysed initially for the inclusion in this collection, the format of the chapter differs from that of the others. The central issues, however, are demonstrably comparable with those of the other contributions.

As Ping-Chao Lee, Alan Bairner and Tien-Chin Tan demonstrate, one of the main objectives of the Beijing Games for the Chinese leadership was to promote the cause of Chinese nationalism. The underlying purpose was to strengthen cohesion within a multi-national and ethnically diverse China and to make the Special Administrative Regions (SARs) (Hong Kong and Macao) identify more strongly with mainland China than perhaps had been the case in the past. In addition, an overt attempt was made to bring Taiwan further into China's ambit by arguing that the use of the name 'Chinese Taipei', insisted on in the past by China to permit Taiwan's Olympic participation and, indeed, its involvement in other transnational activities, should be replaced with 'China Taipei'. Since Hong Kong was referred to for Olympic purposes as 'China Hong Kong', such a name change would have created the false impression of Taiwan as yet another SAR.

As the data presented in this chapter illustrate, although the Beijing Games had the potential to enhance the pride that Taiwan's Han Chinese majority has in its Chinese ancestry and culture, the attempt to translate this pride into closer political identification with mainland China was largely unsuccessful. Once again, however, the Olympics had become embroiled in debates that are fundamentally political. We should not be surprised. As this book consistently reveals, the attempts of Brundage, Pound and many others to keep politics out of the Olympics were always doomed to failure. Indeed, keeping the Olympics out of politics has proved to be just as impossible a task.

CONCLUSION

Despite the best efforts of the IOC and NOCs and their representatives to disguise the issues as discussed in the chapters, and to manufacture consent by instilling in the public the belief that sport in general, and the Olympics in particular, have nothing to do with politics, it is now gradually becoming common knowledge that neither the IOC nor the Games themselves are as apolitical as we have been led to believe. As Triesman (1984: 17) observed, 'all sport is political and the Olympics most political of all'. More and more stories are surfacing about the IOC's abuse of power and position (Lenskyj, 2000; Pound, 2004) and about wrongdoings in the name of the Olympic spirit and Movement (Lenskyj, 2002; Shaw, 2008). Regardless of the emerging evidence, Shaw argues that being critical about and protesting against the Olympics is 'like hating Santa Claus' (Shaw, 2008: xiii). Even in academic circles, in spite of the masses of books written on the Olympics, the narrative often lacks an adequate critical thrust and adopts the celebratory approach of hagiography (for example, as endorsed by Juan Antonio Samaranch, see

Lucas, 1992). In other words, although 'scholars have produced an impressive body of articles and books, they often tend to "toe the party line", and few offer anything that can be considered genuinely critical of any aspect of the Olympic enterprise' (Shaw, 2008: xiv). This cannot be said of the contributors to this collection of essays or of the editors who have written the accompanying glossary. It is hoped that, through our collective efforts, this book offers a fair and accurate account of the relationship between politics and the Olympics.

REFERENCES

Berki, R.N. (1977) *The History of Political Thought: A Short Introduction*. London: Dent.

Entine, J. (1999) *Taboo: Why Black Athletes Dominate Sports and Why We're Afraid to Ask About It*. New York: Public Affairs.

Gibson, O. (2009a) 'Tough task to change the outlook in east London', *Guardian* (London), 27 July, p. 7.

——(2009b) 'Olympics legacy under pressure as sceptics question bold promises', *Guardian* (London), 27 July, pp. 6–7.

——(2009c) 'Funding will be squeezed after 2012 but the hunt for golds is still on track', *Guardian* (London), 27 July, p. 6–7.

Gramsci, A. (1971) *Selections from the Prison Notebooks* (eds G. Nowell Smith and Q. Hoare). London: Lawrence and Wishart.

Guoqi, X. (2008) *Olympic Dreams: China and Sport, 1895–2008*. London: Harvard University Press.

Hampton, J. (2008) *The Austerity Olympics: When the Games Came To London in 1948*. London: Aurum Press.

Hoberman, J. (1997) *Darwin's Athletes: How Sport has Damaged Black America and Preserved the Myth of Race*. Boston, MA: Houghton Mifflin.

Jarvie, G., D.-J. Hwang and M.K. Brennan (2008) *Sport, Revolution and the Beijing Olympics*. Oxford: Berg.

Kent, G. (2008) *Olympic Follies: The Madness and Mayhem of the 1908 London Games: A Cautionary Tale*. London: JR Books.

Lenskyj, H. J. (2000) *Inside the Olympic Industry: Power, Politics, and Activism*. Albany, NY: SUNY.

——(2002) *Best Olympics Ever? Social Impacts of Sydney 2000*. Albany, NY: SUNY.

Lucas, J.A. (1992) *Future of the Olympic Games*. Champaign, IL: Human Kinetics.

Pound, R.W. (2004) *Inside the Olympics*. Toronto: John Wiley and Sons Canada Ltd.

Shaw, C.A. (2008) *Five Ring Circus: Myths and Realities of the Olympic Games*. Gabriola Island: New Society Publisher.

(2009) *Sunday Morning Post* (Hong Kong), Sunday Sport Section, 4 October, 5.

Tomlinson, A. and G. Whannel (eds) (1984) *Five Ring Circus: Money, Power and Politics at the Olympic Games*. London: Pluto Press.

Triesman, D. (1984) 'The Olympic Games as a Political Forum', in A. Tomlinson and G. Whannel (eds), *Five Ring Circus: Money, Power and Politics at the Olympic Games*. London: Pluto Press, 16–29.

Witherspoon, K.B. (2008) *Before the Eyes of the World: Mexico and the 1968 Olympic Games*. DeKalb, IL: Northern Illinois University Press.

Olympic power, Olympic politics: Behind the scenes

HELEN JEFFERSON LENSKYJ

INTRODUCTION

As a secular initiative, the Olympic industry is arguably one of the most powerful as well as one of the most under-scrutinized business enterprises in the world. In critiquing the politics of the Olympic Movement—the International Olympic Committee (IOC) and its subsidiary organizations—I use the term *Olympic industry*. My purpose is to challenge benign-sounding, pseudo-religious concepts such as Olympic Movement, Olympism, Olympic family and Olympic spirit—terms that promote mystique and elitism, while obscuring the power and profit motives that underlie all Olympic-related ventures (Lenskyj, 2000, 2002, 2008). Actual sporting competition represents the mere tip of the iceberg; largely hidden from public view until the bribery crisis of the late 1990s were the operations of IOC, Olympic sponsors, national Olympic committees (NOCs), bid committees, organizing committees for the Olympic Games (OCOGs), and international sporting federations (IFs).

KEEPING POLITICS OUT OF SPORT?

The Olympic Games are by definition political. They involve citizens, politicians and tax payers' money. Furthermore, the IOC demands financial commitment on the part of relevant levels of government as part of the bid process. Those who decry the politicizing of the Olympics are often the individuals and organizations that have the most to lose from protests and boycotts. Interestingly, when politicians and Olympic boosters try to win taxpayers' support (and public funds) for Olympic bids, this is not labelled 'bringing politics into the Olympics'. Nor is it called political when organizing committees lobby politicians to pour unlimited tax money into Olympic spending, as they are currently doing in the upcoming host cities of Vancouver and London. However, when protesters take to the streets to get public attention focused on the misplaced spending priorities in the host city/state/country, or to draw world media attention to local and global injustices, often with considerable success, they are accused of politicizing and contaminating something pure and honourable. 'The eyes of the world' argument promoted by Olympic organizing committees and politicians to justify escalating Olympic budgets, repressive legislation and excessive security has proven useful for activists. Human rights organizations, anti-poverty groups, housing advocates, environmentalists and indigenous peoples have used the opportunity provided by hosting the Games and the accompanying media interest to attract global attention to the injustices that continue in their home countries and internationally.

THE IOC: A PRIVATE CLUB

The IOC resembles a private club in terms of its internal operations and the rules governing eligibility. Although it registered with the United Nations (UN) as an international organization with legal status in 1975, it remains exempt from the provisions of the 1997 Organisation for Economic Co-operation and Development (OECD) Convention on Combating Bribery of Foreign Public Officials.

Countries do not elect representatives to the IOC; rather, in a self-perpetuating process, the IOC appoints men (and a few women) to serve as its ambassadors to member countries. Following a 1966 decision, existing members were entitled to life membership, and the mandatory retirement age for new members was set at 72. Under the presidency of Juan Antonio Samaranch (1980–2001), this was raised to 75, 78, 80, and then back to 70, following the reforms of 1999.

Since 1896, the IOC has had only seven presidents and two acting presidents, namely, 'three barons, two counts, two businessmen, an overt fascist and a fascist sympathizer', as Canadian critic Chris Shaw describes them (Shaw, 2008: 67). In short, the IOC structure virtually guarantees maintenance of the status quo, with members' profound sense of entitlement largely unchallenged.

As early as 1933, IOC President Henri Baillet-Latour exemplified Olympic hubris. Asserting presidential authority over the upcoming 1936 Games in Berlin, he told Adolf Hitler, 'When the five-circled Olympic flag is raised above the stadium, it becomes sacred Olympic ground ... There, I am the master' (cited in Senn, 1999: 53). At the conclusion of the event subsequently dubbed the Nazi Olympics, American IOC member (later IOC President) Avery Brundage pronounced it 'the greatest and most glorious athletic festival ever conducted ... far more than a mere athletic spectacle' (cited in Senn, 1999: 62). More recently, we have witnessed host cities and host countries anxiously awaiting the 'best Olympics ever' presidential pronouncement—the gold medal—during the closing ceremonies; any lesser praise is deemed a failure.

BRIBERY AND CORRUPTION UNCOVERED

The IOC bribery crisis erupted on 9 December 1998, when Swiss member Marc Hodler told an American journalist that it was not legitimate for the Salt Lake City bid committee to have made scholarship payments to IOC members' families (Carter, 1998). In subsequent press interviews, he claimed that the buying and selling of votes had also taken place during the 1996 bid process, with between 5% and 7% of IOC members implicated. He later upgraded that figure to at least 25%, and it was subsequently rumoured that very few members had not been susceptible to bribes. Given the privileged backgrounds of most members, it is difficult to account for the level of greed, or perhaps the level of perceived entitlement, that motivated these individuals to seek and/or accept inducements in exchange for their votes.

An allegation of corruption in the bidding process was made as early as 1987, when reports of a male IOC member demanding sex for votes appeared in a book by Lars Eggertz, chair of the Falun (Sweden) bid committee (Jennings, 1996). IOC executive members Richard Pound and Raymond Gafner dismissed the Falun committee's allegations as the actions of a sore loser. Further allegations of bribery, this time during the 1996 bid process, were published in a German magazine in 1990. Atlanta was a surprise winner, and Athens, considered by many to be the sentimental favourite for the 100-year anniversary of the modern Olympics, a surprise loser. The bid committee representing Athens consulted with other unsuccessful bid committees about their experiences with dishonest IOC members. In January 1991, Samaranch and the IOC executive held a meeting with representatives from these committees. None was prepared to name names even though, in the case of Toronto, full documentation of the abuses was presented.

Hodler's earlier initiatives had forced the IOC to address the vote-buying issue in 1994 by introducing more stringent guidelines, known as Hodler's rules, but the organization would not acknowledge a systemic problem (nor would it do so in 1998). Rather, its rationale for restricting members' travel and expenses was to reduce escalating costs to bid committees (IOC 2000 Commission, 1999). Hodler's rules also established a maximum of US $150 for gifts or benefits given to visiting members and their families. As late as 1999, Pound reiterated the IOC's official position that there was a lack of evidence to support allegations of dishonest conduct.

THE INVESTIGATIONS AND THE CONSEQUENCES

In 1999, Pound was chairing the IOC's Commission charged with bribery, one of several such committees hastily set up in the aftermath of the bribery scandal. Its report, citing the Olympic Charter's wording regarding the offence of acting 'in a way which is unworthy of the IOC', claimed that expelled members were not 'unworthy human beings' but had simply made 'mistakes' (IOC 2000 Commission, 1999: 12, 14). Although it recommended a number of expulsions, this commission under Pound's leadership tended to minimize the offences and to excuse the offenders.

Under both the Swiss Civil Code and the Olympic Charter, expulsion of culpable members was the only possible result, and the report stated that more lenient punishments not provided for in these codes would have been more appropriate in cases of mere 'negligence' or 'carelessness'. Ironically, under Swiss jurisdiction, the IOC apparently had no difficulty in finding an antiquated law that provided grounds for filing a criminal libel suit against UK journalists Vyv Simson and Andrew Jennings, clearly a lesser charge than bribery and corruption. In 1994 the IOC alleged that their 1992 book, *The Lords of the Rings*, defamed Samaranch, an offence that resulted in a five-day suspended sentence and court costs of about $1,000 (Jennings, 1996).

Extensive evidence of bribery and corruption in bidding processes eventually came to light through formal investigations in the USA, Canada, Australia and elsewhere that targeted IOC members, members of bid committees, and the agents and lobbyists who acted on their behalf. Two common defences surfaced in these reports: that the giving of gifts and services simply symbolized *friendship* within the Olympic family, and that such practices were justified because 'everyone was doing it'. Friendship was variously defined as personal relationships with IOC members and their families; the natural outcome of one's own identity as an Olympian and one's cultivation of exclusively Olympian friends. In addition, practices such as gift-giving were deemed customary in the world of international business. It soon became clear that charges of misconduct targeted black or other minority members more often than white members, and that sanctions were less likely to be applied to powerful IOC men, men from powerful countries, or men in the running for IOC president.

MACRO-POLITICS AND OLYMPIC HEGEMONY

These scandals left the Olympic image irrevocably politicized in the public eye, although there had, of course, been political undercurrents shaping the Games ever since their modern revival in 1896. These have been thoroughly documented by Olympic historians and sociologists including Richard Espy (1979), Christopher Hill, (1992), John Hoberman (1986), and Alfred Senn (1999). Most of the earlier crises, however, were at the level of macro politics, with the IOC awarding the Games to repressive regimes in Germany (1936), Mexico (1968), the USSR (1980) and the Republic of Korea (South Korea, 1988) generating widespread international controversy. In all instances, the IOC stood its ground in the face of international pressure and threatened or actual boycotts. Issues of amateurism versus professionalism were also played out at the level of international politics, with the training methods employed in the former Eastern Bloc countries alleged by Western NOCs to constitute state financial support of their athletes, and hence equivalent to professional status.

Sporadic attempts to challenge Olympic hegemony, by organizing alternative international sporting competitions, included the Women's World Games of 1922, the workers' sports movement of the 1920s and 1930s, the Games of Emerging National Forces (Asia, Africa, Latin America and socialist countries) in the 1960s, and the Goodwill Games in 1986. These were largely unsuccessful in the long term. Measured in terms of athlete numbers, however, the Gay Games held in Sydney in 2002, the sixth since their inception in 1982, were larger than the Summer Olympics held in that city two years earlier. However, the Gay Games model gives priority to participation; there are no elimination trials, and most events are not IF-sanctioned. As a result, they do not pose as serious a threat to the Olympic Games as some of the earlier initiatives. Even so, the US Olympic Committee, ever diligent over perceived misuse of Olympic properties, took the organizers to court in the 1980s to prevent them from calling the event the Gay *Olympic* Games (Lenskyj, 2003).

Issues of colonialism and sport are also relevant at the level of macro-politics. UK sport sociologist Richard Giulianotti (2004) examines global sport, colonialism and 'sentimental education' (education that appeals to sentiment rather than reason). He argues against subscribing to 'the more naïve or evangelical arguments regarding sport's *innate* goodness' (emphasis in original) while ignoring 'the historical relationship of sport to forms of colonialism and neo-colonialism' and the ways in which sport entrenches sexism, racism, nationalism and xenophobia (p.367). He further documents how Western sports institutions, including Olympic organizations, have committed acts of 'cultural genocide' not only through their abuse of sport as a colonizing tool but also through 'the deliberate supplanting of non-Western body cultures with imperial games' (p.358).

In promulgating an uncritical view of sport as a social good, the IOC has helped shape UN policy, as well as the programmes of humanitarian non-governmental organizations. As a registered international organization, the IOC was instrumental in the UN General Assembly's 1995 adoption of a resolution titled 'Building a Peaceful and Better World through Sport: The Olympic Ideal'. Congratulating itself on influencing the UN on this issue, the IOC report (1995: 45) noted, 'It is the first time that a non-governmental international organization (the IOC) has enjoyed special treatment within the UN because of the nature of its achievements'. In 2003 a report of the UN Inter-Agency Task Force on Sport for Development and Peace, established in the 1990s, claimed, 'Sport is an ideal school for life ... Sport actively educates young people about ... honesty, fair play, respect for self and others' (UN, 2003: 8).

REFORM EFFORTS

The IOC embarked on a limited programme of reform following recommendations made by the various commissions of inquiry between 1999 and 2000. An internal Ethics Commission was established, limits on age and terms of office were introduced, and membership was expanded to include 15 retired athletes and 15 presidents of IFs.

IF presidents faced a potential conflict of interests because they were likely to focus on promoting their own sports rather the broader goals of the Olympics. Some presidents had a history of putting pressure on bid committees to get the best venues, schedules and accommodations for their own athletes, Sydney's beach volleyball stadium at Bondi being one example. As Olympic scholar John MacAloon noted, IF presidents' involvement promoted cronyism, white elephant venues, unauthorized advantages, and ever-escalating numbers of sports and events (MacAloon, 2001).

Relationships between IFs and the IOC have a history of conflict. During the 1950s state financing of NOCs, most notably those in the Soviet bloc and in non-Western, newly independent countries, had brought with it emerging 'state amateurism', perceived as a serious challenge to the Olympic principle of pure amateurism. The IOC subsequently made rule changes and organized conferences designed to promote co-operation with IFs and NOCs, and to strengthen its hold over these organizations. For their part, most IFs desired sovereignty over their own sports. Not all IFs and NOCs shared the view that the IOC was 'the directive power in the world of sport'; some held that it only had jurisdiction over Olympic sport (Espy, 1979: 51, 53). Some 50 years later, there was little doubt that the IOC did in fact govern virtually all global sports, with the notable exception of soccer, where FIFA's World Cup, first staged in 1930, remains far more significant than the Olympic soccer competition, and tennis in which Grand Slam victories are more eagerly sought than Olympic medals. The introduction of golf to the Olympic programme will represent a similar challenge.

Following the recommendations of the IOC Commission, members were no longer allowed to visit bid cities. This responsibility was now in the hands of an evaluation commission comprising 13 representatives of the IOC, IFs, NOCs, athletes, former OCOG members, and other experts; however, a loophole also permitted IF presidents to visit bid cities in their federation roles.

The election of Jacques Rogge as Samaranch's successor in 2001 was heralded as a symbol of a new era, although it was well known that he enjoyed Samaranch's support (and probably the benefit of his lobbying efforts). One of the three candidates for the presidency, South Korean member Un Yong Kim, had been under investigation in the bribery crisis, and had been issued a 'most severe warning'. When Rogge won a clear majority of votes, Kim alleged that Samaranch had launched a smear campaign against him. None of this deterred the IOC from electing Kim as one of its vice-presidents and an executive member in 2003. However, in 2004, his IOC membership was suspended, and he resigned the following year (rather than facing expulsion) when he was charged with embezzling funds of the World Tae Kwon Do Federation while serving as its head. He subsequently received a two-year jail term in South Korea.

Kim was not the only disgraced international sporting leader. In the first decade of the 21st century, the IAAF (athletics), FIFA (soccer), the International Cricket Council, the Tour de France, and other prominent international sporting organizations were all subjected to unprecedented levels of international scrutiny concerning improper conduct on the part of their leaders and/ or members. More closely linked to the IOC, the patrons of the Seoul 1988

Olympics, Chun Doo Hwan and Roh Tai Woo, were tried eight years later, having been charged with the massacre of hundreds of protesters in 1980. Both were found guilty but their sentences were commuted. Samaranch had publicly praised Roh for his role in promoting democracy in South Korea (Senn, 1999: 286).

Within the 'old boys' network' of long-standing IF leaders, João Havelange (former FIFA President), Joseph Blatter (FIFA President since 1998), and Primo Nebiolo (former IAAF President) are longstanding IOC members. In March and April 2008, during court proceedings in Zug, Switzerland, it was revealed that in the 1990s around £60m. in bribes were paid to sports officials, mostly from FIFA, during Havelange's and Blatter's presidencies; both have also faced other corruption allegations, and at the time of writing, investigations are continuing. With Olympic industry rhetoric focusing on peace-building, intercultural understanding and celebrations of sporting excellence, public attention continues to be deflected from the fact that the IOC harbours a number of sport leaders under suspicion and/or charged with fraud and corruption. Individual men were disciplined, charged, expelled and/or jailed, but there has been little systemic change, and the internal governance of these organizations has undergone minimal reform (Jennings, 2007, 2008; Katwala, 2000).

Women have played a minor role in these developments, being under-represented at the top levels of most IFs; only two of the 35 Olympic sports federations have female presidents. Women are also significantly under-represented in IOC membership, with 16 women out of 110 members, and only one woman, American Anita DeFrantz, serving on the executive board. At the time of the bribery scandal, one lone female IOC member and one of the first two women to be elected in 1981, Pirjo Häggman from Finland, was the first to resign. A later report claimed that she was seeking reinstatement, probably when it became apparent that some male IOC members—Un Yong Kim (South Korea) and John Coates and Phil Coles (Australia), for example—had avoided expulsion despite evidence that implicated them in the bribery scandals.

OCOGS AND THE IOC

OCOGs represent the most visible manifestations of the IOC's operations on the ground, in contrast to the closed-door activities of the organization in Lausanne, Switzerland, or in the various international venues where it holds its annual meetings. An OCOG executive committee, as mandated in the Olympic Charter, must include the IOC member(s) of the country, the president and secretary-general of that country's NOC, at least one member representing the host city, and other representatives of public authorities. While the IOC has 'supreme authority' over the Games, the local OCOG and the host city assume complete financial responsibility, with relevant levels of government backing required to provide financial guarantees. These commercial arrangements distinguish the IOC from a business corporation because 'the buck stops' at the level of its local subsidiaries, not with the IOC itself.

Recent history has shown that the IOC builds not only a financial firewall between itself and OCOGs, but also a moral barrier, with Beijing being a pertinent example. At the time of the bribery scandals, ironically, the firewall offered some protection to OCOGs, whose members could claim that they were not responsible for events in Lausanne. Alternatively, despite losing face by this admission, some claimed they were simply duped by dishonest IOC members. Most often, however, the IOC uses this arms-length relationship to its own advantage, washing its hands of political and social issues in host cities and countries on the grounds that it cannot control local government policies and practices.

THE IOC: LEGAL AND MORAL AUTHORITY

The IOC, like IFs and other business corporations, is legally free to determine its own governance practices. Even national sports federations that choose to be incorporated under state or national legislation have few legal obligations. In Switzerland, the IOC is exempt from federal, cantonal and communal taxes, as well as wealth tax, as a result of an agreement with the Swiss Government signed in 2000 (Shaw, 2008: 72). Furthermore, in 1980, thanks to Samaranch's initiative, the Swiss Government granted the IOC legal immunity. However, although the IOC as a private institution remains immune from government intervention, its locally incorporated subsidiaries, most notably NOCs and OCOGs, are recipients of government funding and subject to state and national legislation.

Examining the IOC as a modern business corporation, legal scholar Saul Fridman (1999) concludes that, although it has the status of a 'legal person' under Swiss law, it has many characteristics that distinguish it from a corporation. He notes that the Olympic Charter has little to say on matters that are typically required of corporations under national incorporation legislation, such as members' and non-members' rights, reporting and auditing requirements, corporate managers' duties and responsibilities, and remedies available to corporate stakeholders. He goes on to develop a persuasive argument to show that the Olympic Movement shares more common ground with the Roman Catholic Church than with a business corporation. Neither is subject to national authorities; members are not democratically elected; both were founded by a small number of individuals and are value-based; and both have faced recent controversies regarding individuals' wrongdoings (and, as more recent history shows, both have failed to undertake systemic change in the face of proven instances of these wrongdoings).

Writing in 1999 at the beginning of the bribery scandal, Fridman noted that both the IOC and the Catholic Church are subject to market forces, and if the IOC 'strays too far from the values enshrined by modern Olympism, the resultant cynicism might well destroy public confidence in the Games and lead to the erosion of the Olympic trademark's value'. Although the IOC's 'straying' reached unprecedented levels, Fridman's prediction of growing public cynicism and reduced sponsorship deals did not prove accurate. Fierce competition continues unabated among bid cities, potential sponsors and television networks for the lucrative Olympic prize, and the era of the 'tarnished rings', to use the mass media's favourite metaphor, was in fact short-lived, thanks largely to the IOC's generously funded public relations machinery and superficial attempts at reform.

From a critic's perspective, the fact that the IOC does not share enough features with corporations to be labelled as one is significant. The Olympic Charter, first compiled in 1978, together with various other pronouncements dating back to 1896 and the emergence of the modern Olympics, clearly demonstrate that the IOC has set itself a higher moral standard than those that apply to a mere international business venture. According to the Charter, the IOC is 'the moral authority for world sport' and the 'supreme authority' over the staging of the Olympic Games and the management of Olympic intellectual property. It is, therefore, legitimate for critics to judge the IOC according to the high moral bar that it has set for itself, and this has been the pattern ever since the 1998 bribery scandal, especially in the mass media and in the court of public opinion. For example, following the leadership role that it assumed in 1999 in relation to doping, with Richard Pound heading the World Anti-Doping Agency, the IOC was justly criticized in international sport circles on the grounds that its own integrity was at issue.

Sports lawyer Alexandre Mestre, a director of the Portuguese Olympic Academy, claims that in sports law, 'there is no text more universal than the Olympic Charter'. Perhaps overstating the case, Mestre asserts that 'It is indeed amazing that a document issued by a Swiss private corporation has assumed all the features of an international treaty!" (Mestre 2007: 7). However, it is true that the IOC has a long history of influencing international politics, as evidenced in controversies over the participation of Taiwan and the People's Republic of China, the USSR, East and West Germany, Israel and the Palestinian Autonomous Areas, South Africa and Rhodesia. As Espy points out, when the IOC recognized a country's Olympic committee, it was in effect 'conferring political recognition although the IOC had no formal diplomatic status' (Espy 1979: 29).

Furthermore, Mestre notes, the Charter has 'an extra-legal aspect—its moral authority, based on the social, economic and sporting significance of the Olympic Games' (Mestre 2007: 7). He cites examples of sport-related legislation in European countries that has been directly shaped by the Charter, and he aptly labels as 'noteworthy' the fact that countries are subject first and foremost to 'Olympic law' when they bid for the Games. Similarly, on the matter of boycotts, there are examples of NOCs defying government policy: for example, the British Olympic Committee's decision to send athletes to Moscow in 1980.

Unlike Mestre, I view these trends from a critical perspective, as disturbing examples of Olympic industry intervention into the domestic politics and legislation of the city, state and country hosting the games. Examples include the following:

- the ban on protest in or near Olympic venues, a requirement spelled out in the Olympic Charter;
- the US Government's quarantine concessions to the IOC in 1996 to allow foreign horses entry to Atlanta;
- the fact that the Olympics Minister in the New South Wales parliament also held the position of head of the Sydney 2000 OCOG, 1996–2000;
- the presence of Israeli and American military personnel on Australian territory, reportedly to protect their countries' athletes, in 2000;
- China's insistence on deploying its own security personnel to monitor the Olympic torch relay overseas in 2008.

BEIJING 2008 AND THE IOC'S 'MORAL AUTHORITY'

It is illuminating to contrast these earlier political interventions with the façade that the IOC maintained vis à vis China in 2008. For example, when challenged regarding China's censorship of websites deemed 'sensitive' (that is, critical of China), the IOC position was that it could not tell the Chinese Government what to do. Significantly, world media attention focused on the website censorship that international journalists would experience, and few Western sources publicized the 21-point list of restrictions that China's propaganda bureau issued to Chinese journalists before the Games (uncovered by *Sydney Morning Herald* journalist Jacquelin Magnay and published on 14 August). The list identified a number of topics that were off limits, including 'all food safety issues, such as cancer-causing mineral water' (Magnay, 2008). In hindsight, this particular edict takes on tragic significance as tens of thousands of Chinese infants became ill after drinking contaminated milk just one month after the Olympics.

A statement by IOC marketing head Gerhard Heiberg on 15 March 2008 during the torch relay protests exemplifies the hypocrisy:

> We still maintain that the Olympics are mainly a sports event and we do not want to get involved in a sovereign state's domestic and foreign policy ... [but] behind the scenes there can be silent diplomacy, trying to explain how things could hurt the success of the games.
>
> (Leicester, 2008)

Significantly, Heiberg relies primarily on a self-interest rationale to persuade China that a negative international image could damage the Beijing 2008 Games, rather than invoking any moral arguments enshrined in the Olympic Charter. Eight years earlier, however, he had taken a values-based approach when interviewed about the bribery scandals and subsequent reforms. Asked about the fact that the IOC still included Blatter and Havelange, members 'allegedly tainted by corruption', he agreed that 'some people' should not be members, but claimed that, with the reforms, 'we are now more protected against the few who put self-interest before that of the movement' (Toft, 2000).

The IOC repeatedly stated that when it awarded the Games to Beijing in 2001, it was with the understanding that China would address the urgent issue of human rights violations. This so-called promise was not, of course, enshrined in the host city agreement, which gives priority to fiscal arrangements and liabilities, and is silent on issues of social justice. However, since references to broken promises on the part of China have been reiterated not only by IOC officials but also by non-governmental organizations like Amnesty International, there probably were informal discussions along these lines. Seven years passed, however, with little or no IOC comment on China's continued record of violations, until protests during the torch relay in March 2008, forced it to make a statement condemning China's military suppression of Tibetan protesters. Even then, with concrete evidence of this violence and calls for boycotts circulating through the global media, the IOC resorted to the old argument that boycotts hurt only the athletes.

Interestingly, in August 2008, Pound told the IOC general assembly that Canada, as well as many other countries, had been 'in boycott mode' at the time of the pro-Tibet protests, but that public opinion (and, indeed, public sympathy) had been affected by the earthquakes in Sichuan (Lee, 2008). For decisions about a national boycott to rest on public opinion is not only disturbing, but also raises serious questions about the professed moral authority of sports leaders.

CONCLUSION

With the lack of accountability and transparency that characterizes the IOC and its subsidiaries, what emerges from this overview of historical and contemporary trends is only a partial picture of the internal political processes. However, it is not difficult to find implicit and explicit evidence of these processes in the words and actions of representatives of the IOC, NOCs, IFs and OCOGs. Overall, despite what might appear to be a worthy goal of serving as a moral authority over world sport, the egregious moral failings of those involved in Olympic organizations make that goal unattainable.

REFERENCES

Associated Press (2008) 'IOC's Rogge says Games boycott would hurt only athletes'. 15 March.

Carter, M. (1998) 'IOC official criticizes SLOC for scholarships'. *Salt Lake Tribune* December 10.

Espy, R. (1979) *The Politics of the Olympic Games*. Berkeley, CA: University of California Press.

Fridman, S. (1999) 'Conflict of interest, accountability and corporate governance: the case of the IOC and SOCOG'. *University of New South Wales Law Journal*, 22:3 www.austlii.edu.au/au/journals/UNSWLJ/1999/28.html.

Giulianotti, R. (2004) 'Human rights, globalization and sentimental education: the case of sport'. *Sport in Society*, 7(3): 355–69.

Hill, C. (1992) *Olympic Politics*. Manchester: Manchester University Press.

Hoberman, J. (1986) *The Olympic Crisis: Sport, Politics, and the Moral Order.* New Rochelle, NY: Aristide D. Caratzas.

IOC (1995) *Citius Altius Fortius,* 3:1, 45.

IOC 2000 Commission (1999) *Report by the IOC 2000 Commission to the 110th IOC Session,* Lausanne, December 11 and 12. Lausanne: IOC.

Jennings, A. (1996) *The New Lords of the Rings: Olympic Corruption & How to Buy Gold Medals.* New York: Pocket Books.

——(2007) *Foul!: The secret world of FIFA: Bribes, Vote Rigging and Ticket Scandals.* London: HarperSport.

——(2008) 'FIFA 'misled' detective on trail of missing pounds paid for World Cup TV rights'. *Telegraph,* 30 July, www.telegraph.co.uk.

Katwala, S. (2000) *Democratising Global Sport.* London: Foreign Policy Centre.

Lee, J. (2008) *Earthquake changed Canada's mind on Olympic boycott – Pound. Vancouver Sun,* 5 August, www.canada.com/vancouversun/news/sports/beijing2008/story.html?id=c74b839a-98dd-4112-9c55-b084a2e20127.

Leicester, J. (2008) 'Calls mount for Olympic ceremony boycott'. *Associated Press* 18 March, apnews.myway.com/article/20080318/D8VG3QDGO.html.

Lenskyj, H. (2003) *Out on the Field: Gender, Sport and Sexualities.* Toronto: Women's Press.

——(2008) *Olympic Industry Resistance: Challenging Olympic Power and Propaganda.* Albany, NY: SUNY Press.

——(2002) *The Best Ever Olympics: Social Impacts of Sydney 2000.* Albany, NY: SUNY Press.

——(2000) *Inside the Olympic Industry: Power, Politics and Activism.* Albany, NY: SUNY Press.

MacAloon, J. (2001) *IOC Reform Then and Now: An Insider's View.* Presentation, University of Toronto, 5 February.

Magnay, J. (2008) 'Censors make news'. *Sydney Morning Herald,* 14 August, www.smh.com.au.

Mestre, A. (2007) 'The legal basis of the Olympic Charter'. *World Sports Law Report* November, 6–7.

Senn, A. (1999) *Power, Politics, and the Olympic Games.* Champaign, IL: Human Kinetics.

Shaw, C. (2008) *Five Ring Circus.* Vancouver: New Society Press.

Toft, J. (2000) 'Continued need for IOC reform'. *Play the Game,* www.playthegame.org/uploac/12–13-playthegame.pdf.

UN (2003) *Sport for Development and Peace.* New York: United Nations.

Wamsley, K. (2004) 'Laying Olympism to rest', in J. Bale and M. Christensen (eds), *Post Olympism? Questioning Sport in the Twenty-first Century.* London: Berg Publishers, 231–42.

FURTHER READING

Horne, J. and W. Manzenreiter (eds) (2006) *Sports Mega-Events.* Oxford: Blackwell.

Jennings, A. (with Clare Sambrook) (2000) *The Great Olympic Swindle: When the World Wanted its Games Back.* London: Simon & Schuster.

Schaffer, K. and S. Smith (eds) (2000) *The Olympics at the Millennium: Power, Politics and the Games.* Piscataway, NJ: Rutgers University Press.

The politics of hosting the Olympic Games

JOHN HORNE

INTRODUCTION

'In this city, you were either working for the Olympics, or you were dreading them—there was no middle ground.'
(Manuel Vázquez Montalbán (2004/1991) *An Olympic Death.*
London: Serpent's Tail, 34)

Occasions such as the Olympic Games, the football World Cup and other sports mega-events, act as socio-cultural reference points, and reveal both the appeal and elusiveness of sport. In the age of global television, moreover, the capacity of major sports events to shape and project images of the host city or nation, both domestically and globally, makes them a highly attractive instrument for political and economic elites. It is in this context that the pursuit of hosting sports mega-events has become an increasingly popular strategy of governments, corporations and civic 'boosters' world-wide, who argue that major economic, developmental, political and socio-cultural benefits will flow from them, easily justifying the costs and risks involved (Horne and Manzenreiter, 2006). Numerous studies fuel the popular belief that sport has a positive impact on the local community and the regional economy. Sport has been seen as a generator of national and local economic and social development. Economically it has been viewed as an industry around which cities can devise urban regeneration strategies. Socially it has been viewed as a tool for the development of urban communities, and the reduction of social exclusion and crime.

Compared with this conventional—or dominant—view of the Olympic Games and the Olympic Movement, here are a series of conclusions derived from a recent book and documentary film about the 2010 Vancouver Winter Olympics (Shaw, 2008; Schmidt, 2007). The Olympics can be seen as a tool used by business corporations and governments (local, regional and sometimes national) to develop areas of cities or the countryside. They permit corporate land grabs by developers. Five major construction projects have taken place in association with the 2010 Winter Games—the building of the Canada Line (formerly known as the RAV—Richmond Airport Vancouver—Line) connecting the airport and downtown Vancouver, the athletes' village and a convention centre; and developments in the Callaghan Valley west of Whistler (the main skiing area where the Olympic snow sport events will take place), and the building of an extension to the 'Sea to Sky Highway' through Eagleridge Bluffs, in West Vancouver, to enable faster automobile transportation between Vancouver and Whistler. The view of one of the contributors to the film was that it was a disaster for any city on the planet to host the Olympics. Host city populations face increased taxes to pay for the 'party'. The poor and the homeless face criminalization and/or eviction as downtown areas are gentrified—improved to appeal to more affluent visitors or full-time residents. The hosting of such a mega-event skews all other economic and social priorities and means the loss of the opportunity to do other things with public resources spent on the Games. The IOC markets sport as a product, pays no taxes and demands full compliance with its exacting terms and conditions, including governmental guarantees about meeting financial shortfalls. The end results are 'fat-cat' projects and media spectacles benefiting mostly the corporations that sponsor the Games, property developers that receive public subsidies, and the IOC which secures millions of dollars from television corporations and global sponsors.

Clearly when considering the politics of the Olympic Games, the role and impartiality of the researcher is called into question—as Montalbán suggests in the quotation above (he was originally writing just before Barcelona hosted the Olympics in 1992), researchers may find that there is no middle ground. Amongst the questions that this chapter will attempt to answer, none the less, are: Why do governments and cities compete for the right to host major international sporting events such as the Olympics? What are the commercial underpinnings of hosting the Olympic Games? What can we learn from recent 'bidding wars' about the contemporary politics of the Olympics? How do Olympic 'boosters' and 'sceptics' portray the 'legacies', economic and otherwise, which are proclaimed for the Games?

TO BID OR NOT TO BID? IS THAT A SERIOUS QUESTION?

The attraction of hosting an Olympic Games (or another sports mega-event) is possibly greater today than ever before. It is widely assumed that hosting a Games represents an extraordinary economic opportunity for host cities (and nations), justifying the investment of large sums of public money. Yet this view has only developed in the past 30 years. Hosting the 1976 Summer Olympics resulted in huge losses and debts for the city of Montréal. The debt incurred on the interest for the loans to build what turned out to be largely 'white elephant' sports infrastructure was only finally paid off in November 2006—costing the Montréal taxpayers well over C $2,000 m. in capital and interest costs, without anything like commensurate benefits. Rather than experiencing a post-Olympic boom, the economy of Montréal in the mid-1970s went into a steep decline that would last for almost two decades. No wonder, then, that when the Los Angeles Olympics took place in 1984 there had been no other city seriously bidding to host the event (Whitson and Horne, 2006).

Today, established cities in advanced capitalist societies and cities in developing economies alike weigh up the possibility of hosting the Olympic Games. At the time of writing the hosts for the next three Olympic Games (2010, 2012 and 2014) are known and four cities are about to submit 'bid books' for 2016. At any one time, then, at least three cities are anticipating hosting an Olympics and others are waiting their turn. The change in the allure of hosting the Olympics has partly come about because of the success that the LA Games appeared to represent—in terms of making a substantial financial surplus or profit of over US $200 m., laying a solid economic foundation for a support system for athletes in the USA, and putting on a television spectacular involving many of the world's athletes. The attraction of hosting has also come about because of changes that the International Olympic Committee (IOC) has made to the process of selecting cities following investigative journalists' revelations of insider corruption in the 1980s and 1990s (see Jennings and Simson, 1992; www.transparencyinsport.org).

Two examples illustrate this change in the seriousness with which bidding is treated and the attraction of the Olympics. First, today the IOC establishes an Evaluation Commission (EC) to visit Candidate Cities bidding to host the Olympic and Paralympic Games—both events now taken as part of the Olympic host 'package'. Composed of representatives from the Olympic Movement as well as a number of advisers, the EC analyses the candidature files of cities bidding to host the Olympics after they have been submitted to the IOC. It then makes on-site inspections and publishes an appraisal for IOC members before they meet to elect a host city. An illustration of the seriousness with which governments and politicians approach hosting the Olympics is the increased involvement of leading politicians in the process. Although scheduled to host a summit of the G8 leading economic nations in Gleneagles in Scotland, the then UK Prime Minister, Tony Blair flew to Singapore in July 2005 in order to meet IOC members and representatives ahead of the crucial vote deciding which city would host the 2012 Olympics. When the IOC voted in 2007 that the Winter Olympic Games of 2014 would take place in the Russian resort of Sochi, the then Russian President, Vladimir Putin, performed a similar role.

The major inducement to engage in Olympic hosting now as opposed to in the 1980s is, of course, financial—the sponsorship and television rights money that the IOC has negotiated largely covers most of the *operating* costs of the Olympic Games—between $2,000m. and 2,500m. for the Summer Olympics. The Games attract vast television audiences—the 2008 Beijing Olympics, for example, drew an estimated cumulative global television audience of 4,700m. over the 17 days of competition, according to market research firm Nielsen. Such estimated audience figures have to be treated with caution, but Nielsen's estimate surpassed the 3,900m. viewers for the Athens Games in 2004 and the 3,600m. who watched the 2000 event in Sydney. The 2008 Beijing Olympics was also the most-viewed event in US television history; according to Nielsen, 211m. viewers watched the first 16 days of Olympic coverage on US network NBC.

Large television audiences have meant that television corporations and broadcasting unions have been prepared to pay increasing sums of money to the IOC for exclusive coverage of the Olympics, which has helped them to offset the operational costs of the Games. Hence, in future, alongside cities such as Paris, London, New York, Chicago, Madrid and Beijing, it is likely that smaller cities, for example Copenhagen (Denmark) and those in the 'global south', such as Durban (South Africa) and Delhi (India), will consider bidding for the Summer Olympic Games in 2020 and beyond. Acting as host to the IOC's Congress (Copenhagen 2009, Durban 2011) or one of the 'lower order' sports mega-events (Delhi will host the Commonwealth Games in 2010) is considered as a way of leveraging support for a subsequent Olympics bid.

COMMERCIAL CONSIDERATIONS

Commercially today in addition to broadcast partnerships, the IOC manages the TOP (The Olympic Partner) world-wide sponsorship programme and the IOC official supplier and licensing programme. Since 1985, when the TOP programme started, the financial health of the IOC has been secured by the first two sources—television rights payments and global sponsorship deals. As an article in *The Economist* put it ahead of the Atlanta Summer Olympics in 1996, the 'zillion dollar games' have developed because 'the power of corporate hype linked with global television is a marvellous machine for promoting sports'. Television rights account for slightly more than 53% of IOC revenue and it is likely that television income will continue to increase. Income for the 2010 and 2012 Games is already assured at $3,800m., an increase of 40% on the $2,600m. the IOC received for the 2006 and 2008 Games. At the time of writing, negotiations are being conducted for the 2014 and 2016 events, and with such large viewing figures from Beijing, the fees are likely to increase substantially irrespective of the location of the Games.

The IOC refers to its financial operations in terms of an 'Olympic quadrennium'—a four-year period (1 January—31 December). The latest one for which accounts have been made available (2001–04) generated a total of more than $4,000m. in revenue. The IOC distributes approximately 92% of Olympic marketing revenue to organizations throughout the Olympic Movement to support the staging of the Olympic Games and to promote the world-wide development of sport, and retains the rest to cover operational and administrative costs of governing the Olympic Movement. The IOC provides TOP programme contributions and Olympic broadcast revenue to the Organizing Committees of the Olympic Games (OCOGs) to support the staging of the Games. Long-term broadcast and sponsorship programmes enable the IOC to provide the majority of the OCOGs' operational budgets well in advance of the Games, with revenues effectively guaranteed prior to the selection of the host city. The two OCOGs of each Olympic quadrennium share approximately 50% of TOP programme revenue and value-in-kind contributions, with approximately 30% provided to the summer OCOG and 20% provided to the winter OCOG. During the 2001–04 Olympic quadrennium, for example, the Salt Lake 2002 Organizing Committee received $443m. in broadcast revenue from the IOC, and the Athens 2004 Organizing Committee received $732m. The OCOGs in turn generate substantial revenue from the domestic marketing programmes that they manage within their host countries, including domestic sponsorship, ticketing and licensing. National Olympic Committees (NOCs)—of which there are over 200—receive financial support for the training and development of Olympic teams and athletes. The IOC distributes TOP programme revenue to each of the NOCs throughout the world. The IOC provided approximately $318.5m. to NOCs for the 2001–04 quadrennium.

Although there appear to have been many positive developments since the Los Angeles Olympics, several academics and sports people at the time (see the contributors to Tomlinson and Whannel, 1984) and since have been critical of the increasing commercialization of the games and the likely impact this has had on the event. Portrayed as the views of 'gloom merchants' and 'naysayers' by those involved with the Olympics and associated sports federations, these criticisms are not in fact simply voiced by people who want to put an end to the Olympics. With the increasing involvement of powerful global brands as Olympic sponsors has come attendant commercial rights legislation—to provide exclusivity to their association with the Olympic symbols (the interlocked rings, the name of the games, etc.) and avoid 'ambush marketing' which the corporations pay millions of dollars to obtain. Yet this is seen as overly restrictive by smaller businesses and organizations. The Olympic Games also provides a major attraction to sponsors at a national level and thus drains resources away from other non-Olympic sports and cultural activities during the build-up to the event. Criticisms of the IOC as an organization have also had some impact on its practices.

The IOC remains a private organization which only accepts invited members. The voting membership of the IOC currently consists of slightly more than 110 people, including the President, Jacques Rogge, of whom only around 15 are women and active athletes. The IOC contains several members of royal families and corporate leaders and people holding an executive or senior leadership position within an IF or an NOC. It thus remains subject to accusations of lack of transparency whilst claiming to be a movement and a 'family' based on a philosophy beyond politics. Alongside the myths and ideology of Olympism—with elements such as the creed, the motto and the flame borrowed from Christianity (Catholicism and Protestantism)—it is not surprising that quasi-religious claims are often made such as upholding the 'spirit' of the Games. Critics prefer to portray the Olympics nowadays as an 'industry', a 'machine' and even a 'disease' that creates a blight on the cities and their populations that act as its hosts. These discursive differences manifest themselves especially in the politics of hosting, and the public relations wars to which we now turn.

'PR WARS'

Mega-events such as the Olympic Games provide multiple meanings for different groups of people—as they happen, when they have taken place and, perhaps especially, as they are being bid for. Hence, we know that advocates of hosting the Olympics will deploy a range of discursive strategies to win over public opinion. The main issues around which the hosting of the Olympics has been debated involve the burden of the costs and the distribution of the benefits. Research points to the uneven impacts of the Olympics. Despite much media acclamation, and the accolade 'the best games ever' being proclaimed at the closing ceremony by the outgoing IOC President Juan-Antonio Samaranch, the 2000 Olympics in Sydney generated substantial negative impacts on local residents and the environment—giving evidence to the claim that there is potential for conflict between economic and social benefits realized from hosting sports events. Since the late 1970s (and the Montréal Olympics in particular), a major concern in considerations of the Olympics has been this gap between the forecast and actual impacts on economy, society and culture. That there is likely to be such a gap is now fairly predictable. Pro-hosting advocates tend to gather and project optimistic estimates whilst anti-hosting groups articulate concerns. More generally, there has been an overestimation of the benefits and an underestimation of the costs of mega events.

The positive achievements claimed by Games boosters include increased employment, a boost for tourism, opportunities for civil engagement (through volunteering), and emulation in terms of increased active involvement in physical activity. In addition, there is a 'trickle-down' assumption that industries and other parts of the host city's nation will benefit from the economic upturn and demand for goods and services stimulated by hosting the Olympics. Certainly recent past and selected future Olympic hosts have made these arguments. One of the main problems regarding the assessment of the costs and benefits of mega-events relates to the quality of data obtained from impact analyses. Economic impact studies often claim to show that the investment of public money is worthwhile in the light of the economic activity generated by having professional sports teams or mega-events in cities. Yet here much depends on predictions of expenditure by sports-related tourism. Research shows that many positive studies have often been methodologically flawed and the real economic benefit of such visitor spending is often well below that specified. According to the European Tour Operators Association (in *Olympics and Tourism: Update on Olympics Report 2006*), the Olympic Games are 'an abnormality that is profoundly disruptive' of normal patterns of tourism. Another measure of economic impact—on the creation of new jobs in the local economy—has often been politically driven to justify the expenditure on new facilities and hence the results are equally questionable.

With respect to social regeneration, it has been noted that there is an absence of systematic and robust empirical evidence about the social impacts of sports-related projects. Some research suggests that there may be positive impacts from greater community visibility, enhanced community image, the stimulation of other economic development and increases in 'psychic income'—collective morale, pride and confidence. Most commentators appear to agree that there will be a positive outcome with respect to health promotion, crime reduction, education and employment and general 'social inclusion', but without actually having the evidence to support that view. The problem is that there have been very few research projects in this area and the methodologies needed to investigate these issues have not yet been adequately developed.

Since the early 1990s, when investigative reporting by journalists and social researchers uncovered details of corruption in the Olympic Movement, and such news began to damage the reputation of the IOC, the organization and OCOGs have engaged public relations (PR) companies and spin doctors to assist in managing media messages and the global and local image of the Olympics. News and image management, spin doctoring and PR have become key features in any major public policy development both in the United Kingdom and throughout the rest of the world.

Clearly the mass media are centrally important in discussing PR 'wars'. Are the media boosters or sceptics? It depends. The private sector media—in the United Kingdom the newspaper press and independent television, for example—could be critical if it suited their interests. The public sector—especially the BBC in the United Kingdom—has tended to help to sell the 2012 bid and the associated hosting to an uncertain public. The BBC provided saturation coverage of the 2008 Beijing Olympics sending over 440 journalists and reporters—an unprecedented number. In 2008 the BBC could have been accused of indulging in much 'jock sniffing'—adoration that leans towards the idolatry of professional and elite athletes. This in turn creates an expectant audience. Irrespective of occasional critical comments and blogs by journalists, the BBC is and will remain a major booster for the London Games. Contradictory opinion poll findings after the Beijing Games about public confidence in the London 2012 Organising Committee (LOCOG) team, however, suggest that a considerable number of Britons remain to be convinced that the 2012 Games will be a success. Some populations of course are resistant to the allure of the Games. In September 2008, 51% of Norwegians were reported as being against giving state guarantees of between NOK 15,000m. and 20,000m. ($4,000m.–$5,000m.) for Tromso to host the 2018 Winter Olympic Games. A month later, the city had withdrawn from the bidding process altogether. The IOC expects to see evidence of public support in any country that applies to host the Olympics.

Just as reputation and symbolic power have become increasingly valuable resources for elected politicians, so too are they vital for International Sports Organizations and International Sports Federations (IFs). In this environment 'crisis communications' in response to bad publicity during 'spin wars' have become part of the PR role. It has been argued that the PR job is essentially to secure or 'manufacture' the consent of the public, which covers both active support and passive acquiescence for economic and social policies and developments. More broadly, Miller and Dinan (2007: 13) argue that the PR role has become 'to position private interests as being the same as public interests' and, in so doing, to undermine the meaning of a public interest separate from that of private corporations. In this respect, PR is as much concerned with bringing other private businesses and civic leaders 'onside' as members of the general public. As it has developed, a PR company's task is often to predict and thus ward off damaging attacks, especially in debates in the public sphere about urban development, such as the hosting of the Olympics. Hence, an advertisement in *The Guardian* from the London 2012 Olympic Delivery Authority (ODA) asked for 'Community Relations Executives' who would act as:

> the main link between the ODA project teams, contractors and local residents with special regard to the construction impact and have the ability to win the trust of sceptical local audiences through strong interpersonal, influencing and communications skills.
>
> (*The Guardian*, 24 March 2007, Work section: 12)

OLYMPIC BOOSTERS AND OLYMPIC SCEPTICS

Richard Cashman (2002: 5) suggests that there are four periods during which the issues and questions surrounding specific Olympic Games and Paralympic Games are keenly debated: the preparation of a bid and winning the right to play host; the long seven-year period as hosting is prepared in the successful city/nation; the relatively much shorter staging of the Games; and after the games have taken place. I suggest that we should expand this typology—there are at least eight periods surrounding each Olympics—and the range of issues and questions Cashman identifies. A more appropriate time period for the analysis of all the issues and debates surrounding the hosting of an Olympic Games might be closer to 20 years. Such a long time frame is necessary to capture the different moments when boosting and criticizing take place. During these phases debate and controversy can surround: the decision to bid, the extent and nature of community consultation, the issues of impacts and opportunity costs and benefits, the involvement of anti-Olympic and community lobbies, the impact of public order and Olympic-related legislation on human rights, the scale of the event, as well as more generally the relation of the Olympic Games to housing, the environment and not to forget sport, in both elite and mass participation terms.

When London gained the 2012 Olympic Games in July 2005, the coverage in the London-based British national press and the London local press was broadly celebratory. Newspapers in Scotland, however, were more questioning, raising concerns about the potential negative impact on the public funding of events in Scotland and the distribution of benefits from a London Games. In the six months prior to the Beijing Olympics, the issue of escalating costs, which had come to dominate the domestic news agenda about London, took second place to the issues surrounding Beijing—the torch relay and protests about human rights and Tibetan independence especially. Then as the period of London's Olympiad began seriously public relations events were staged to recreate the positive feeling stimulated by the successful podium position of 'Team GB' in Beijing (such as the 'parade of heroes' held in London in October 2008), launch 2012-related events (such as the start of the Cultural Olympiad at Tate Britain's Duveen Galleries in September 2008), and alert the London and British public more generally to new developments and milestones in the build-up to 2012.

To fully understand the patterns of politics and protest during these phases, it is thus necessary to consider the specific local and regional as well as national and international political environments and the politics of sport and sport politics at these levels. Hence, in the case of Sydney, and felt throughout Australia generally in the 1990s, was the issue of Aboriginal rights and reconciliation. This underlay much of the build-up to the event, and during the Games in 2000 the performances on the track of Cathy Freeman were interpreted as part of the ideological battleground where these issues continued to be fought out. Beijing in 2008 was surrounded by debates about the rights and wrongs of awarding the Games to a city in a regime where democratic freedoms and human rights enjoyed elsewhere have been consistently absent. Prior to the bid decision (in 2003), those hoping to put forward London as a credible candidate for the 2012 Summer Olympics had to face substantial lingering concerns about the capability of the British organizing team to deliver such an event and build the necessary facilities. This was in the light of the failure to host the 2005 World Athletics Championship even after being selected, the costly Millennium Dome (now renamed the O_2 Arena) and the spiralling cost of rebuilding Wembley Stadium. With respect to international politics, so long as the United Kingdom remains one of the two main armies fighting in Iraq, London features as a serious target for further terrorist attacks. It is not so surprising, therefore, that Mike Lee, often referred to as 'the spinmeister', who made the transition from political adviser to former Home Secretary David Blunkett, was responsible for orchestrating the London 2012 Bid. After the successful bid, Godric Smith, former adviser to Tony Blair, replaced him as head of communications for LOCOG.

At any Olympics there is a range of boosters and sceptics involved across various fields—sport, politics, business, local community, arts and culture, academia and the media. While it would be incorrect to say that these form homogeneous groups—non-Olympic sports and organizations interested in mass participation, for example, would not be as positive in their comments on the development of a sports mega-event for elite athletes in pursuit of the medal podium if it meant that resources were taken away from them—it is evident that boosters possess considerable power and influence whereas sceptics often feel that they are taking on an impossible task. Sceptics have fewer resources, lack easy access to the mass media, and often lack credibility. The 'chilly climate for Olympic critics' that Lenskyj (2002) identifies extends to all sceptics and groups formed to protest against public policies and the social and economic status quo. Specifically, objecting to a sports mega-event has the additional problem of making politics out of something that is portrayed as socially valuable and/or fun. As a member of the No Games 2010 Coalition in Vancouver said, 'going up against the Games ... was like going up against Santa Claus. Why would you do that?'

Partly as a means of facing up to this challenge, Olympic watchdog groups often adopt irony and humour as a tactic in self-naming. Hence the Canadian based group, Bread Not Circuses (BNC), formed in the 1990s to challenge the Toronto Summer Olympics proposal, while PISSOFF (People Ingeniously Subverting the Sydney Olympic Farce) and CACTUS (the Campaign Against Corporate Tyranny with Unity and Solidarity) were established in Sydney before the 2000 Olympics. However, there are divisions amongst the sceptics as well and monitoring was the main focus of the IOCC—Impact on the Community Coalition—in Vancouver. During the plebiscite held in 2002 about whether the bid for the Winter Olympic Games should go ahead, the IOCC took a position in favour of the Games when the campaign boosters had met enough of their concerns. Shaw (2008) argues that this was an example of the use of opposition groups as social safety valves, diverting serious opposition and thus serving the interests of the bid promoters.

The power imbalance between boosters and sceptics appears unquestionable; in terms of resources (financial, temporal, as well as strategic), access to media sources, intermediaries and credibility, boosters appear unassailable. None the less, public scepticism about boosting can be considered a source of potential contestation that challenges the hegemony. The basis for a successful 'anti' campaign requires the establishment of an alternative common sense, effective mobilization of competing discourses, and the creation of a different visual ideology of urban space. Specific issues, such as pollution—noise, crowds, inconvenience—and the inappropriateness—on environmental or other grounds—of the chosen location, are likely to be mobilized when those affected view themselves as citizens with a common purpose rather than isolated consumers, and begin to question the meaning of their 'civic duty'. Organizing an effective and successful 'anti' campaign is very difficult to achieve but this is not an argument against it being attempted.

Shaw (2008) outlines different forms of resistance to the Olympics. Where opponents are vocal, united and strong, they have derailed bids to host the Games before they have got off the ground, for example in Amsterdam (1992), Melbourne (1996), Finland (2006), Berlin (2000) and Toronto (in 1996 and 2008). Lack of public support (as noted in Norway in 2008) is not attractive to the IOC. If some additional risk—greater than usual—is associated with a potential host then opposition coalitions stand a greater chance of success. Where there is little unity, or internal division, groups are less effective in stopping bids, such as in Atlanta (1996), Sydney (2000), Salt Lake City (2002) and Vancouver (2010). After successful bids, some opposition coalitions have been able to continue on a small scale, as in Calgary (1988), Nagano (1998) and Torino (2006). Only one anti-Games campaign has succeeded after the Games were awarded, in Denver in 1976. Following the award of the Winter Games for 1976, a public referendum voted against proceeding with the hosting on the basis of the additional local tax burden and environmental impact (Innsbruck in Austria replaced Denver as the 1976 Winter Olympic Games hosts). With respect to Vancouver 2010, the next Winter Olympics, in addition to the book by Shaw (2008), a film of the same name by Conrad Schmidt (2007) and an alternative mascot have been produced to try to win over some hearts and minds.

The strategies of critics of the Olympics vary depending on the view held of both the IOC as an organization and the local Olympic Games promoters. If the Olympics are seen as a front for real estate development and projects largely benefiting private corporations with substantial help from public funds, then the tactics may include the creation of an international anti-Olympic movement along the same lines as the anti-globalization movement, making association with the event, through volunteering for example, seem 'uncool', and taking direct action such as creating protests that spoil the spectacle. On the other hand, if the contemporary Olympics are seen as a distortion of a fundamentally sound vehicle for the promotion of international peace through sport and cultural activity, then constructive engagement with the organizations that make up the Movement—the OCOGs, the NOCs and the IOC itself—might be attempted. This is the position of groups such as Amnesty International and the Centre on Housing Rights and Evictions (COHRE), which have been critical of IOC decisions about hosts but have also attempted to engage the organization in dialogue. In this respect, these groups are closer to the approach of the alter-globalization movement that is underpinned by the idea that 'another world is possible'.

'THE IMAGINEERING OF LEGACIES'

Another way of illustrating these differences of tactics is to consider different uses and framing of the word 'legacy'. Legacy is a very positive word, implying the passing on from one generation to the next of something of substantial benefit. Another word, which critics would argue often refers to the same thing, is 'cost'. This is value laden with a more negative nuance. For this very reason from a social scientific point of view, a better word to use is possibly 'outcome', as it is more objective, less value laden. Yet, 'legacies' are now very much part of the IOC and Olympic bid rhetoric, especially when associated with regeneration—of urban infrastructure, of populations and of place marketing. The media help to 'imagineer' images of host locations as 'world class'—for people inside as well as outside the host city and country (see Rutheiser, 1996 for a detailed account of the Atlanta Games). There are three main propositions in what Shaw (2008) calls the 'Olympic framing' of legacies.

First, it is suggested that the Olympics will make money for the host city—the economic proposition. Second, the Olympic Games is portrayed as a 'once in a lifetime' opportunity for the city and nation—the patriotic proposition. Third, the Olympic Games are portrayed as an event for athletes, sport and the Olympic ideals (grace, beauty, youth and international goodwill)—the idealistic proposition. In reality the first proposes a gamble, the second refers to an intangible, and therefore hard to contest, experience, and the third appeals directly to a population's philanthropy—making rejection appear cruel and unkind. Before a bid to host the Olympics, all three of these propositions feature widely. After a successful bid, though, the first (economic) proposition is downplayed, and as the event or bid decision approaches, any problems that arise can be blamed on critics and sceptics because of their negative attitudes.

If we substitute legacy for costs, a different set of issues emerges. There are four main sets of costs, but only two of these are usually regarded as Games-related. The foremost economic analyst of the Olympics, Holger Preuss (2004), and the IOC tend to focus attention on the *operating* costs of an Olympics. Much of this is now met by money from the TOP programme and broadcasting rights revenues. In addition there are *venue construction* costs, *infrastructure* costs and *security* costs. Venue construction costs, like operational costs, are often downplayed even though they are paid for out of public money. Infrastructural changes, however, are often regarded as though they were going to happen anyway, albeit accelerated by the Games, and are sometimes claimed as part of the Olympic legacy, even though they are not included in the calculation of costs to the public. 'Stratford City', a retail and shopping mall development being built by Australian property developers Westfield in East London provides a good example of this. In September 2008 David Higgins, the LOCOG Chief Executive, claimed in a London 2012 blog that the Olympics were responsible for a major retail development in the locality. Yet the decision to develop Stratford actually took place *before* the Olympic Games were awarded to London. Now called 'Stratford 2011' the rebuilding project was approved in 2004 before the bid was won, and was designed to stand alone even if the Olympics did not come to London. Little of this development had anything to do with the Olympics specifically, but is now included in the list of Olympic Games achievements. Thus, a legacy, which should come after the passing of an event, is claimed in advance by boosters. Finally, the costs for security have been growing exponentially since 2001 and these are predominantly borne at public expense. Increased security measures also add to the social costs of the Games in terms of increased surveillance of public spaces and the general militarization of the Olympic host city and venues.

CONCLUSION

Writing this during an apparent crisis of capitalist financial regulation, in which vast sums of public money are being used to support institutions floundering owing to the poor judgements and speculative deals of private financiers, is a useful reminder that government intervention to support advanced capitalism is not an aberration. The Olympic Games surely provide a positive, uplifting, even inspiring diversion during times like these, and arguably there are more serious topics for a social scientist to research. However, as this collection will make clear, while the Olympic Games is a global spectacle attracting vast audiences, it is also a deeply political phenomenon. Thus, researching the costs and benefits of sports mega-events is itself a highly political activity. It can involve researching the powerful—the elite of business leaders and government officials—as well as leaders of sports organizations who gather together to formulate hosting bids or stage events after a successful bid has been made. It can also involve research into the ongoing activities of organizations more sceptical of the benefits of the event that are themselves investigating the claims made by mega-event promoters or 'boosters'.

The IOC claims that the Games offer inspiration, as a Movement, a family and a philosophy. Because the Olympic Movement is derived in many respects from Pierre de Coubertin's Christian beliefs and his relationship to organized religion of the late 19th and early 20th centuries, being critical of the Games can be likened to 'farting loudly during High Mass in the Vatican' (Shaw, 2008: 154). Yet, the Olympic Movement creates attention for itself by inviting criticism when it deviates from its own proclaimed ethical standards. The gap between rhetorical claims and actual outcomes is one means of attracting charges of hypocrisy. Developments in capitalism over the past 25 years provide the background against which the allure of hosting the Olympic Games has grown. Yet it also creates the conditions in which criticism of the Games develops. As Shaw (2008: 168) states, 'In our consumer culture, mega events, Olympics included, use sports as a platform to sell stuff'. The negative 'legacies' associated with the Games include social polarization, eviction and displacement of marginal populations, public resources being used for private benefit, global rather than local benefit and the creation of playgrounds for the affluent. When normal political routes and legal processes are short-circuited by governments and local organizers anxious to get work completed on time, it is not surprising that critics consider tactics that they hope may create the biggest public relations impact.

REFERENCES

Cashman, R. (2002) *Impact of the Games on Olympic Host Cities*. Barcelona: Centre d'Estudis Olimpica (UAB).

Horne, J. and W. Manzenreiter (eds) (2006) *Sports Mega-Events*. Oxford: Blackwell.

Jennings, A. (with Clare Sambrook) (2000) *The Great Olympic Swindle: When the World Wanted its Games Back*. London: Simon & Schuster.

Lenskyj, H. (2008) *Olympic Industry Resistance: Challenging Olympic Power and Propaganda*. Albany, NY: State University of New York Press.

——(2002). *The Best Olympics Ever? Social Impacts of Sydney 2000*. Albany, NY: State University of New York Press.

Miller, D. and W. Dinan (2007) 'Public relations and the subversion of democracy', in W. Dinan and D. Miller (eds), *Thinker, Faker, Spinner, Spy*. London: Pluto, 11–20.

Preuss, H. (2004) *The Economics of Staging the Olympics: A Comparison of the Games 1972–2008*. Cheltenham: Edward Elgar.

Rutheiser, C. (1996) *Imagineering Atlanta: The Politics of Place in the City of Dreams*. London: Verso.

Schmidt, C. (2007) *Five Ring Circus: The Untold Story of the Vancouver 2010 Games*. Canada: Ragtag Productions, documentary film available from www.thefiveringcircus. com.

Shaw, C. (2008) *Five Ring Circus: Myths and Realities of the Olympic Games*. Gabriola Island: New Society Publishers.

Tomlinson, A. and G. Whannel (eds) (1984) *Five Ring Circus: Money, Power and Politics at the Olympic Games*. London: Pluto Press.

Whitson, D. and J. Horne (2006) 'Underestimated costs and overestimated benefits? Comparing the outcomes of sports mega-events in Canada and Japan', in J. Horne and W. Manzenreiter (eds), *Sports Mega-Events*. Oxford: Blackwell, 73–89.

FURTHER READING

Close, P., D. Askew and X. Xu (2006) *The Beijing Olympiad: The Political Economy of a Sporting Mega-Event*. London: Routledge.

Jennings, A. and V. Simson (1992) *The Lords of the Rings: Power, Money & Drugs in the Modern Olympics*. London: Simon & Schuster.

Roche, M. (2000) *Mega-events and Modernity: Olympics and Expos in the Growth of Global Culture*. London: Routledge.

Politics of gender through the Olympics: The changing nature of women's involvement in the Olympics

SUSAN J. BANDY

INTRODUCTION

In many countries contemporary sports have been considered a means towards women's emancipation with the female athlete as the best exponent of female liberation, and the Olympic Games are seen as a 'marker' or signifier of this emancipatory process. The global significance of the Games has brought increasing attention to the female athlete as women throughout the world have sought parity in all aspects of cultural life. In this regard, the Olympics have stimulated debates over such issues as female athleticism, equality of opportunity in sport, the status of women in cultures throughout the world, human rights and the sex/gender distinction, and the unfairness of sex testing of female athletes.

Levels of participation of women in the 112 years of the modern Games have increased from zero in the 1896 Games to 4,746 female participants in 127 events in the 2008 Games in Beijing. This increase in the number of participants and events, however, masks the overall ongoing inequality of male/female participation in the Games. There is still a gap in the participation rates of males and females, with female athletes comprising only 42% of the total number of Olympians and participating in only 127 events (42.1%), while men competed in 165 events (54.6%) in Beijing (in addition to 10 mixed events). Countries continue to send teams to the Games without female representation. Such disparities can be seen as indicative of gender inequality throughout the Olympic Movement, a form of inequality which is even more visible at the organizational and administrative levels. The Olympic Movement and the International Olympic Committee (IOC) continue to be dominated by men, and women are vastly under-represented on the IOC, the international sports federations (IF), and National Olympic Committees (NOC).

The participation of women in the Olympic Games and their struggle for equality within the Olympic Movement have been widely studied. Their historic under-representation in the Games and the various obstacles women have faced have been well researched and documented. The history of the battle for acceptance and recognition in sport in general, and in the Olympic Games in particular, has been marked by struggles over the control of international sport for women, male domination and female subordination in sport, and inequitable treatment of female athletes. The history of the Olympic Games and the Olympic Movement suggests that the early years of the creation of the IOC and the establishment of the Games actually extended the unfair treatment of female athletes and the discriminatory treatment of leaders of international sport in the early years of the 20th century resulting in battles over the control of women's sport during the formative years of the Olympic Games as well as on-going struggles for admittance to the Games and equitable treatment of women in the Olympic Movement throughout most of the 20th century. An analysis of the early years of the IOC, the pattern of development of the participation of female athletes in the Games, the evolution of the Olympic sports programme, and the gradual admittance of women in the IOC, the NOCs and other international governing bodies in sport reveals the changing nature of women's involvement in the Olympic Games.

DE COUBERTIN AND THE EARLY YEARS OF THE IOC AND THE OLYMPIC MOVEMENT

An understanding of the nature and function of the IOC is important and instructive in comprehending the manner in which the programme of the Games would be developed and the control of women's participation in international sport would be secured. In 1894 Pierre de Coubertin organized a congress to promote the development of amateur sport with which the 'gentlemen' of the early 20th century were associated, with the intention of ensuring the regular celebration of the Games, worthy of high amateur ideals, while simultaneously promoting and strengthening friendships between *sportsmen* in all countries. Members of this group were socially prominent men, the majority of whom came from upper-class European society. The elitist and private nature of the Committee was revealed in the original statement, which described the organization as a permanent, self-perpetuating body 'selecting for life such members as it considers qualified' (cited in Mitchell, 1977: 210). Therefore, entrance into the organization which controlled the Olympics was restricted to individuals who shared the same basic philosophy of sport. Further political autonomy of the group was established from the onset; members did not represent their countries on the IOC but were instead representatives of the IOC to their countries (Schneider, 2000). The intention was to create an organization devoid of political influence with the Committee governing as a whole in the interest of amateur sport. The Olympic Movement was thus conceived as a tool to promote and spread European aristocratic and masculine values. The criteria for joining and composition of the membership ensured that the IOC could develop an Olympic programme according to its own wishes and remain uninfluenced by opinion outside the sphere of the gentleman amateur. The Committee was male-dominated, self-regulating and functioned in an autocratic fashion. Moreover, the male domination of the IOC, its political autonomy and insularity, and its intentions to guard and perpetuate amateur sport for men inevitably excluded women from the IOC, the Olympic Movement and the Olympic Games for a good part of the 20th century as all decisions regarding women's participation were to be ultimately controlled by the IOC.

The views and actions of de Coubertin reveal the basis of the opposition to the inclusion of women in the Olympic Games and attempts to gain control of women's sport at an international level during the first three decades of the 20th century. De Coubertin was the primary opponent of women's involvement, and he used his influence as President of the IOC to exclude women from participating in the Games. Although he was unable to prevent their participation entirely, he was certainly able to limit it and his views have been largely perpetuated by successive IOC presidents until the presidency of Juan Antonio Samaranch, which began in 1980.

It has been noted that de Coubertin was indeed 'a man of his times'—indicating, of course, that his ideas concerning women and their participation in sport were consistent with those of aristocratic men of 19th-century Europe (Leigh, 1974: 24). At the time of the establishment of the Games, women's roles—at least in Europe, the British Empire and North America—were changing dramatically with the industrialization and modernization of society and the impact of social reform through the women's movements at the end of the 19th century as women became more physically active. Although de Coubertin supported the importance of physical activity for women, his views concerning women and sport were determined by the 'laws of nature', which supported the notion that women were not suited for strenuous sports competition (Leigh, 1974: 19). The popular view among physicians and many leaders of sport at the time maintained that women were physiologically, anatomically and psychologically inferior to men and, thus, incapable of participating in strenuous physical activity and sport. In addition, social attitudes concerning the private, social role of women, as wives and mothers, further precluded them from the public arena of sport that required independence, competitiveness and strength, each of which was antithetical to the ideals of femininity at the time. De Coubertin's views on women's participation in sport logically led to his position concerning the question of the participation of women in the Games and his disapproval of the indecency, ugliness and impropriety of women in particular sports. He was adamantly opposed to women's participation in track and field as these events were thought not only to encourage an unwomanly competitiveness but were also considered injurious to women's health. Moreover, de Coubertin thought that it was indecent for spectators to see women's bodies in athletic competition (Leigh, 1974).

De Coubertin used women's exclusion from the Ancient Olympics in Greece to justify their exclusion from the modern Games: 'as no women participated in the Ancient Games, there "obviously" was to be no place for them in the modern ones' (Simri, 1979: 12). Furthermore, in the *Olympic Review* of 1912, he defined the Games as 'the solemn and periodic exaltation of male athleticism, with internationalism as a base, loyalty as a means, art for its setting, and female applause as reward' (DeFrantz, 1997: 18). The role of women was to be glorified in the number and achievements of her children, her greatest accomplishment being to encourage her sons to excel rather than to seek records for herself (Leigh, 1974).

As mentioned earlier, however, de Coubertin was unsuccessful in his attempts to completely exclude women from the Games during his tenure as president of the IOC. Although only men competed in the first modern Games in 1896, women would gradually enter the Games and have competed in every Olympiad since 1900.

WOMEN'S ENTRANCE INTO THE GAMES AND THE STRUGGLE FOR CONTROL OF WOMEN'S INTERNATIONAL SPORT: 1896 TO 1936

With the success of the 1896 Olympics, the IOC accepted the obligation of maintaining the programme of the Games. It was at this time that the IOC began to develop the framework for governing the programme and establishing the features which maintained the autonomous position of the organization within international sport. During the first three Olympiads of 1896, 1900 and 1904, the power to organize the Games in Athens, Paris and St Louis, respectively, was delegated to the organizing committees of the host cities and was, therefore, not controlled by the IOC. The host committees were allowed to determine, to a large extent, the events that were included in the programme of the Games. As a result, women were first admitted to the Games without the official approval or consent of the IOC. In 1900 women made their debut in golf and lawn tennis. In 1904 they entered tennis and archery, and, in 1908, they participated in skating, tennis, archery and in demonstrations of gymnastics and aquatics. In 1910, the International Swimming Federation (FINA) voted to include women in the swimming competitions for the Olympic Games in Stockholm in 1912. The organizing committee in Stockholm, with the backing of FINA, added swimming and diving to the competitive programme for women. This suggests that the organizing committees were not opposed to women's competition. Nevertheless, the nature of the sports chosen by each committee—those that were socially sanctioned at the time—suggests that competition only in certain types of sports was acceptable.

It would be easy to assume that the IOC possessed very little control over the Olympic Movement at this time. However, scholars have asserted that this is an erroneous assumption. De Coubertin was held in high esteem by the members of the IOC, was for many years the sole financial supporter of the IOC, and held complete power within the movement for 31 years (Mitchell, 1977).

By 1912, the IOC was gradually becoming more involved in the programme, although no major policies were established at this time. Immediately following the Games of 1912, the IOC convened with the specific purpose of establishing a set programme of Olympic competition. The committee was now prepared to assert its authority and actually control the development of the programme without external influences. De Coubertin was determined to prevent women from having open access to the competitive programme, but his determination did not bring about the withdrawal of the women's events that he had proposed. It did, however, enable the IOC and IAAF to prevent the 'contamination' of the Games by the inclusion of female runners, jumpers and throwers.

The control of the programme by the IOC was critical in terms of its effect on the development of the programme and, thus, on female participation. Other decisions concerning a feasible programme of the Games—including the elimination of tennis and minor women's sports, support for the principles of amateurism and individualism (thereby excluding team sports), and a redefinition of internationalism to refer to sports practised by a minimum of six countries—further inhibited the participation of women in the Games (Mitchell, 1977).

Changes in the larger context of international sport would provide the framework for struggles over the control of women's sports that soon followed the decisions taken by the IOC in 1912. While the IOC attempted to secure its control over the Olympic programme and, thereby, exclude women from athletics and inhibit the development of competition for women with the Games, a number of developments external to the IOC would eventually lead to battles for control of women's international sport.

International sports federations and national Olympic committees began to form, and these generally had a more favourable view of women's competition than that of de Coubertin and the IOC (Wilson, 1996). By 1912, a number of international federations (including the IAAF and federations for gymnastics, fencing, football, sailing, rowing and weightlifting) had been established, and research shows that the existence and appearance of an international federation at the Games corresponded to the acceptance of women on the programme (as evidenced by the inclusion of swimming and diving for women and FINA in 1912).

Contemporaneous with the increasing opposition of de Coubertin and the IOC was the establishment of organizations for sporting women in Europe and North America as a direct consequence of the refusal of the male establishment to concern itself with the interest and needs of sporting women. In France, a club known as *Fémina-Sport*, which combined sport and dance, was formed in 1911 and this was followed by the founding of *Academia* which staged the first athletics meeting for women in 1915. In December 1917 the leaders of this club (all men) created the first national women's sports federation, *Fédération des Societés Féminines Sportives de France* (FSFSF). Alice Milliat, a rower and a member of *Fémina-Sport*, was elected as treasurer. By

1918, Milliat had become the General Secretary, and was chosen as President the following year (Leigh and Bonin, 1974). In 1919 the FSFSF organized championships in field hockey, association football, basketball and swimming. In the USA, meanwhile, the American Physical Education Association formed a committee on women's athletics (1917) and, in 1922, the National Section for Women's Athletics was created (Dyer, 1982). German women formed the *Berliner Sport-Club* and national championships for German women were inaugurated in 1920 (Guttmann, 1991). In 1918 the Austrian Amateur Association held national championships for women and the following year, the United Kingdom officially recognized the inclusion of a 440-yard relay for teams in Inter-Services Championships and then at the women's AAA event. In 1920 England hosted the first women's international match in soccer, and a year later Austria, Belgium, Czechoslovakia, France, Finland, Germany and the Netherlands hosted national championships for women (Dyer, 1982).

As these federations were founded and international competitions began in several sports, a growing number of female sporting stars attracted the attention of the media. These included the international tennis stars, Suzanne Lenglen of France and Helen Wills of the USA, the Norwegian skater Sonja Heni, the German airplane pilot Elli Beinhorn and the American swimmer Gertrude Ederle. Media interest would continue into the 1930s stimulated by the achievements of well-known Olympians such as Fanny Blankers-Koen (Netherlands), Mildred 'Babe' Didrikson (USA) and Stella Walsh (Poland).

The combination of these forces revealed a burgeoning interest in international sport for women and inspired Milliat, the President of FSFSF, to propose the inclusion of track and field in the programme of the Olympic Games in 1920. De Coubertin's reaction was to suggest that all events for women be removed from the programme. As Milliat recounted, she 'came up against a solid wall of refusal which led to the creation of the Women's Olympic Games' (cited in Carpentier and Lefèvre, 2006: 1,114). At the IOC's 1920 session in Antwerp, de Coubertin exhibited his continuing opposition to women in the Games and called for the suppression of participation of women in all events, suggesting that it was illegal.

In response to the reaction of de Coubertin and in defiance of the IOC and the IAAF, Milliat organized a women's international athletics competition in Monte Carlo in March, 1921. Five nations (England, France, Italy, Norway and Switzerland) participated in the competition. A subsequent international match was held between France and England in October of the same year. At this time, Milliat decided to form an international organization for women that would provide opportunities for competition as well as regulation. On 31 October 1921, at the inaugural congress in Paris (with representatives from Austria, Czechoslovakia, England, Germany, Scandinavia, Spain and the USA) the *Fédération Sportive Féminine Internationale* (FSFI) was formed and began to function like the IAAF in promoting and governing international sports for women (Leigh and Bonin, 1974). The intention of the FSFI was to recognize international competition, establish records and organize the

Ladies' Olympic Games, scheduled for the third year of the Olympiad. A second athletics competition was organized in Monte Carlo in April 1922 and at the second congress of the FSFI, the members decided to organize the Women's Olympic Games that were held in Paris later in 1922—a one-day event in which five nations took part in a programme that consisted of 11 events (six more than the five allowed by the IOC when it admitted women to track and field competitions in 1928).

The IOC began to take an interest in women's international competitions and, in 1923, suggested to the international federations that they take control of women's activities. The council of the IAAF drew up rules for women's track and field in 1924 and adopted rule changes permitting the inclusion of women. However, the IAAF promptly denied women the right to participate in track and field at the Olympic Games of 1924. Women renewed their application for the Games of 1928 and, although IOC President Count Henri de Baillet-Latour of Belgium, tried to prevent the inclusion of a women's track and field programme too, he was defeated by a majority vote (Dyer, 1982).

The IOC could no longer ignore the controversy that was developing in relation to track and field competitions, and the question of the legitimacy of female athletics was, therefore, the subject of their Pedagogic Conference in 1924. The medical evidence, considered to be 'authoritative' and 'scientific', was presented and later published in a paper entitled 'Women's Participation in Athletics'. Part of the report focused on differences between male and female athletes, the limitations of female biology and, in particular, the potentially harmful effects of athletics on women's reproductive potential (Hargreaves, 1994).

Dismayed by the rapid expansion of the women's programme of the FSFI and the international interest in women's athletics, the IAAF expressed its concerns and then established a special commission on women that initiated a series of negotiations with the FSFI. Sigried Edström, President of the IAAF who later became President of the IOC, cleverly proposed the inclusion of track and field for women in the programme of the Games of 1928, a move that eventually gave the IAAF control of international athletics for women (Carpentier and Lefèvre, 2006).

In 1926 the IAAF finally agreed to include five athletics events for women at the 1928 Games (instead of 11, which had already been contested in the Women's World Games). In return, Milliat and the FSFI agreed to drop the word 'Olympic' from the name of the women's games and accepted the invitation. However, Milliat decided to continue the World Games and subsequent events were held in Gothenburg (1926), Prague (1930) and London (1934). These competitions were unpredictably successful, attracting large numbers of competitors and spectators from Western Europe, British colonies and dominions, and North America, establishing athletic events for women on an international scale. In the face of hostility from the leaders of the IOC and IAAF, women had created an autonomous realm of sport which they could control and develop (Hargreaves, 1994).

Table 3.1 Women's participation in the Olympic Games under the presidency of de Coubertin (1896–1924)

Year/ location	Women's participation	Men's participation	Women's sports
1900 Paris	22	975	Golf Equestrian sports (mixed) Tennis (mixed) Sailing (mixed)
1904 St Louis	6	645	Archery
1908 London	37	1,971	Figure skating (mixed)
1912 Stockholm	48	2,359	Swimming Diving Equestrian sports (mixed)
1920 Antwerp	65	2,561	Swimming Figure skating Diving Equestrian sports Tennis Sailing (mixed)
1924 Paris	135	2,954	Fencing Swimming Diving Equestrian sports Sailing (mixed)

The number of participants and the details concerning women's sports can be found on the official website of the IOC

In 1928 some 277 women participated in five events (100 metres, 800 metres, 4 x 100 metres relay, high jump and discus) in the Games in Amsterdam, more than three times the number who participated in the Women's Games in 1926. Although this was a great success for women, the now infamous 800 metres race, in which some of the women struggled across the finishing line, raised numerous objections to the place of women in the athletic events. Newspaper accounts reported a growing sentiment that competition was too strenuous and that women should be prevented from participating in 'inappropriate' activities. Female athletes were referred to as 'Amazons' and 'masculine' by some writers, while others noted women's 'inferior' muscular power and emotional 'instability' (Spears, 1976: 60). The publicity surrounding the controversy reopened the 19th-century debate about the desirability of women's sport and opponents of women's participation used reports of this race as justification for removing the 800 metres race from the programme. Indeed, for the next 32 years, there was no women's race longer than 200 metres in the Olympic track and field programme.

In the two decades following the Games of 1928, women made few gains. Progress was hindered by the interruption caused by the Second World War;

the attitudes of some female sports leaders and physical educators who opposed elite level sport for women, particularly in the USA; the belief that women were not physically suited for strenuous competition in athletics; and the widespread view that women should be prohibited from sports (Wilson, 1996). Leaders of both the IOC and the IAAF held similar views to those of de Coubertin concerning the participation of women in sport. Count Henri Baillet-Latour, who became President of the IOC in 1925, suggested that women participate in 'aesthetical' events such as gymnastics, swimming, skating and tennis (Dyer, 1982: 123). Avery Brundage (President of the Amateur Athletic Union of the United States, President of the United States Olympic Committee, and later member and President of the IOC), who was elected to the IAAF's 'Committee on Women's Sports', argued that competitive games should be for 'normal feminine girls and not monstrosities' (letter to Baillet-Latour, cited in Leigh, 1980: 16). In 1930 Olympic Congress members again discussed the question of female participation and rejected attempts to include athletics and fencing. In 1934 the IAAF and FSFI established a joint committee and, by 1935, the FSFI was so displeased with the handling of the women's programme that it requested that women be excluded from participation. Each year Milliat and her organization had asked for a full programme of events and each year their petition was rejected.

In 1936 the IAAF took complete control of track and field and ended its co-operation with Milliat and FSFI. The most significant development during this period was the increased number of participants in women's athletics. In 1936 some 331 women from more than 20 countries competed in the Games in Berlin. However, in spite of the number of participants, the programme remained minimal as leaders of the IAAF and the IOC remained resistant to the involvement of women in track and field. After the Games of 1936, Brundage remarked that he was ' … fed up to the ears with women as track and field competitors … [their] charms sink to less than zero. As swimmers and divers, girls are beautiful and adroit, as they are ineffective and unpleasing on the track' (cited in Simri, 1979: 38). He continued to oppose the participation of women and, in 1949, as President of the IAAF, he argued that the track and field programme should be removed from the Olympics. He averred, 'I think women's events should be confined to those appropriate for women—swimming, tennis, figure-skating, and fencing, but not shot-putting' (cited in Simri, 1979: 38).

To gain entry into the Games of 1936, women gave up control of their own athletic competitions. They also accepted a greatly reduced programme which reflected what men in the controlling sport association thought women could and should do, and not what women had already proven they could accomplish. In its 10-year struggle with the IOC and the IAAF, however, the FSFI had played a crucial role in the development of women's sport within the Olympic Movement, primarily by putting increased pressure for gender equality on the IOC and the IAAF. Milliat and the FSFI provided the impetus for many women's international meetings held in the 1920s and

1930s. Moreover, the FSNI was the principal organization for women's athletics until 1936, when it relinquished full control of international women's athletics to the IAAF.

DEVELOPMENTS AFTER THE SECOND WORLD WAR

The Second World War temporarily halted the development of women's sport; however, the war greatly enhanced the emancipation of women in many aspects of life, including sport. Although the 1950s was a conservative era during which traditional ideas concerning women's roles were reasserted and discussions concerning the appropriateness of strenuous athletic competition for women reappeared, in the aftermath of the Second World War, women made significant gains in sport. A record number of 385 female athletes from 33 countries participated in the 1948 Olympic Games.

After the War, the international federations continued to exert pressure on the IOC to include more women's events in the Games (Wilson, 1996; Chase, 1992). In 1953 the IOC passed an amendment aimed at continuing to limit the number and type of sports in which women could participate: 'Women are not to be excluded from the Games, but only participation in "suitable" sports' (Chase, 1992: 35). The interweaving of sport with politics and the efforts of individual countries to accumulate Olympic medals, however, provided a major incentive to improve the standards of elite sports for women and make women's events integral to national sports polices. With the entry of the USSR (and Eastern bloc countries), further pressure was exerted. International politics transformed the Olympics into a battlefield where superpowers flexed their muscles to prove their own ideological superiority, with female athletes becoming pawns in the struggle for power and influence (Wushanley, 2000). In 1955 the USSR—presumably in an attempt to win more Olympic medals—proposed the addition of basketball, rowing, speed skating and volleyball. In addition, in 1957, a proposal was made to 'allow women to compete in the sports included in the Programme of the Games according to the Rules of the International Federations which practice women's events'; in other words, to include women's competitions in any sport that held a recognized women's world championship (Wilson, 1996: 185). Later, in 1964, at the 62nd session in Tokyo, the IOC approved a proposal for a detailed study of the Olympic Programme, with a view to overcoming certain shortcomings as regards the participation of women.

During the 1960s and 1970s the IOC allowed the programme for women to increase, with the addition of the following sports: archery, basketball, luge, rowing, shooting, speed skating, team handball and volleyball. These decades also witnessed modest increased representation of women in NOCs and in the international sports federations.

The emphasis on female athletes as potential medal winners precipitated another controversial aspect of the Games. As early as 1936, Avery Brundage had questioned the sex of some of the female athletes, and as a result of

Table 3.2 New women's sports on the Olympic Programme

Year	Sport	Year	Sport
1900	Tennis, golf	1972	Archery
1904	Archery	1976	Rowing, basketball, handball
1908	Tennis	1980	Hockey
1912	Swimming	1984	Shooting, cycling
1924	Fencing, ice skating	1988	Tennis, table tennis, yachting
1928	Athletics, gymnastics (teams)	1992	Badminton, judo, biathlon
1936	Skiing	1996	Football, softball
1948	Canoeing	1998	Curling, ice hockey
1952	Equestrian sports	2000	Weightlifting, pentathlon, triathlon
1960	Speed skating	2002	Bobsled
1964	Volleyball, luge		

enhanced performances by female athletes from Eastern European socialist countries in the 1960s (most particularly the emergence of East German women who dominated athletics and swimming), the issue reappeared. The sex test was introduced in 1966 at the European Athletics Championships in Budapest following allegations that some female contestants were male (Pittaway, 1999). The Medical Commission subsequently initiated a femininity control programme in 1968. The Olympic authorities instituted a chromosome test for all women competitors in Mexico City, and while none were disqualified, several recent Eastern European champions failed to show up for the competition. It was argued that these tests were a means to preserve the integrity of the Games for female participants, to protect women from being forced to compete against males in disguise and to prevent genetically 'anomalous' individuals from competing (Wackwitz, 2003). Tests continued until 1998 and have remained controversial, with their critics arguing that when women challenge traditional notions of 'femininity' by becoming seriously competitive athletes, they call into question the permanence and immutability of their assigned sex category.

Despite the increase in the number of female participants and events for women, the IOC remained an exclusively all-male organization throughout these decades. However, in 1973, after the resignation of Avery Brundage, the Committee voted to change its rules so that women could be elected to that body for the first time. The election of Ireland's Lord Killanin as the President of the IOC marked a turning point concerning the inclusion of women in the governance and structures of Olympic sport. In the 1968 session in Mexico City, a proposal to add women to the IOC was made, although it would be another 13 years before the Committee actually allowed women to become members. Although Killanin has been regarded as sympathetic to the inclusion of women in the IOC, he failed to achieve much in this respect, remarking that 'We will not elect a woman just because she is a woman' (cited in Leigh, 1980: 19).

GRADUAL ADVANCES TOWARD GENDER EQUALITY

Second-wave feminism and demands for more equitable treatment in sport and other spheres of life such as politics, religion and education corresponded with gradual advances toward gender equality in sport. Public awareness of gender issues influenced the IOC to make changes in relation to the inclusion of women in the Games and the Olympic Movement. After Juan Antonio Samaranch was elected as President of the IOC in 1980, there was a significant, arguably radical, reconstruction of Olympic traditions, rules and regulations. This reconstruction increased the number of female participants in the Games and within the IOC, NOCs and IFs. In 1982 Pirjo Haggman of Finland and Flor Isave Fonseco of Venezuela were chosen to be the first women elected to the IOC. In the same year, Mary Alison Glen-Haig of the United Kingdom was also elected. Samaranch referred to their election as a 'revolution' that should 'incite all other members of the Olympic family to consider the question of women's participation and ... form a powerful invitation to those IFs which have not yet given a place to female administrators' (Samaranch, 1981: 696). Within the next five years, three more women were added to the IOC.

In the 1990s, there were a number of IOC initiatives aimed at addressing matters relating to gender equality in the IOC and the Olympic Movement as the lack of representation of women had become readily apparent. In 1991 the IOC amended its charter to add 'sex' when referring to discrimination within the Olympic Movement so as to read: 'Any form of discrimination with regard to a country or a person on grounds of race, religion, politics, sex, or otherwise is incompatible with belonging to the Olympic movement'.

With the increased participation of women in the Games, the globalization of sport and the force of the women's movement, female leaders in sport began to create international organizations to further the development of sport for women. In 1994 an international conference devoted to Women's Sport and the Challenge of Change, was held in Brighton in the United Kingdom. From this came the formation of the International Working Group (IWG) and the Brighton Declaration, which was addressed to governments, federations and educational institutions and called for the full involvement of women in every aspect of sport.

Shortly thereafter direct, external pressure was placed on the IOC early in 1995 when the Atlanta Plus Committee was formed in Paris by Linda Weil-Curiel, a French activist and lawyer, Annie Sugier, an activist who worked in French nuclear protection, and Anne Marie Lizim, a former Minster of European Affairs in Belgium (Holder, 1996). In 1996 this committee called attention to the fact that at the 1992 Summer Olympics in Barcelona, 34 countries had no female athletes. In 1992 the largest all-male delegations were Iran with 40, Qatar with 31, Pakistan with 27 and Kuwait with 36. Of the 9,959 athletes registered as participating in the 1992 Summer Games, 2,851 were women—fewer than 30%. The group demanded that the IOC refuse to allow those countries without female athletes to participate in the Olympic Games in Atlanta. Basing their demands on the IOC's own Olympic Charter, members of the group argued that if South Africa could be barred from the Games for practising racial discrimination, then other nations should be sanctioned for sex discrimination. The initial response from the IOC was that the exclusion of women from teams representing Muslim countries was not a concern of the Committee and tried to discredit the group as being anti-Islamic (Wilson, 1996). The issue has yet to be fully resolved as nation states still send teams without female athletes to the Games. Nevertheless, the work of Atlanta Plus has further challenged the IOC to recognize its discrimination against women.

At approximately the same time as the formation of Atlanta Plus, Islamic women themselves began to organize sport for women. The First Islamic Countries' Sports Solidarity Congress for Women took place in Tehran in October of 1991, during which the Islamic Countries Women's Sport Solidarity Council was formed. In February 1993 the first Islamic Countries' Sports Solidarity Games were held in Tehran and these have been contested every four years since. The support that the IOC has given to these Games has been challenged by Atlanta Plus, with the view that support of these games will not advance the cause of sport for Muslim women, not least in the Olympics.

The low representation of women in leadership positions was discussed at the 1994 Centennial Olympic Congress, and recommendations were made to alleviate the problem. These included increasing membership of women in the IOC and the commissions within the IOC, and increasing the number of women's sports and disciplines on the Olympic programme. The extent of male domination was also evident in the IFs and the NOCs. In 1995 only two of the 34 IFs representing sports on the Olympic programme were headed by women, and of 196 NOCs, only five had a female president. At this time, Samaranch created the Working Group on Women and Sport to develop a strategy for implementing the recommendations of the Centennial Congress and the group became the Women and Sport Commission in 2004. At present, 42% of the members are male.

The Working Group developed a strategy for increasing the number of women in leadership positions. All NOCs, IFs and national federations were strongly 'invited' to increase the number of women in all of their decision-making structures to at least 10% of the total by 31 December 2000 and reach 20% by 31 December 2005 (Wilson, 1996). Furthermore, the Executive Board agreed that the Olympic Charter should be amended to take into account the need to maintain equity between men and women. It was understood that the recommendations would also apply to the IOC itself. To that end, more women were elected to the Committee and, in 1996, their number had increased to seven out of 106 members and, by 2001, to 13 out of 122. The number has continued to increase under the leadership of Samaranch's successor, Jacques Rogge. Since 1981, a total of 23 women have served as IOC members and there are, at present, 16 women out of 107 active members. In addition, more than 30% of the NOCs and 29% of the IFs have already achieved the target of 20% of female members, according to the claims of the IOC (www.olympic.org, 2009).

While women have yet to become equal participants in the Olympic Games and the Olympic Movement, their equal right to participate has been largely established as a goal within the movement and there has been steady progress towards their integration and equal representation in the Games as well as in leadership positions in international and national sport. Ongoing criticism of the Games continues, however, and focuses on several issues including the failure of media coverage in terms of gender parity, the 'masculinist' nature of the core values and traditions of the Games, and recent developments concerning transgender and transsexual athletes.

As efforts were made to increase the participation of women in the Games and in leadership positions in IFs and NOCs, scholars noted the disparity in media coverage of female athletes in the Games. Despite claims that the 1996 Summer Games in Atlanta were 'the Olympics of the Women', analysis of the coverage showed no significant gains in parity for women (Eastman and Billings, 1999). Ongoing investigations suggest that disparity continues with regard to the scale and character of the coverage of female Olympians by the media.

Critics have also maintained that even though women have been more integrated into the Olympic Games, their inclusion has not altered the core values and traditions of the Olympics. According to Burstyn (1998), these remain masculine in nature with sport viewed as a 'zero-sum contest' of cultures that worship competition, domination and hyper-masculine qualities which, in turn, serve governments, social policy and the political use of sport for nationalist purposes. Further, the Olympic Movement, the IOC and the Games have most frequently been politicized at the level of the nation state with contending political ideologies while transnational corporations use them for commercial purposes (Burstyn, 1998). As Hargreaves (1994: 234) noted, women move into the 'existing patterns of participation and structures of control' rather than attempting to make qualitative changes in the governance structures of the IOC, the NOCs and the international federations.

The IOC has more recently been criticized for its policies concerning transgender and transsexual athletes. Under the Stockholm Consensus, the IOC implemented a policy for the participation of transgender and transsexual athletes in May 2004. The IOC Medical Commission proposed that transsexual athletes who have had Sex Reassignment Surgery (SRS) before puberty will be admitted to the competition. All other transsexuals must be post-operative, must have legal and governmental recognition from their country of citizenship, hormonal therapy to minimize 'gender-related advantages' in competition, and must have lived for a minimum of two years in their newly assigned gender (Cavanagh and Sykes, 2006: 75–76). Scholars have argued that the IOC's policies concerning transsexuals and transgender athletes, as in the case of sex testing, do not reflect an acceptance of gender variance in the world of sport. Rather, there is on-going resistance to inclusive gender policies in mainstream sport organizations. Moreover, the IOC continues to define and regulate what constitutes a female athlete and who is eligible for women's sport (Teetzel, 2006). The IOC's resistance is 'based on anxieties about the instability of the male/female gender binary and the emergence of queer gender subjectivities within women's, gay, and mainstream sporting communities' (Sykes, 2006: 3). Critics claim that such policies are used to further institutionalize sex-gender distinctions and protect the stability of these socially-defined categories and systems of sexual division. In so doing, binary sex-gender systems require that women maintain a lesser status than men, both in and out of competition. Thus, the system which seeks to provide a space for women to compete is also the system that insists upon their competitive inferiority (Wackwitz, 2003).

CONCLUSION

The history of the participation of women in the Olympic Games reflects a pattern of gradual acceptance of the female athlete. However, the IOC did so reluctantly and fought to control this participation from the early years of the Games, wresting control of women's international sport from female leaders in 1936 and contributing to the demise of the FSFI. It was not until 1996, in the face of pressure from international groups such as the IWG and Atlanta Plus, that the IOC began to make changes to increase the participation of women in its governance and administrative structures. Critics argue that these structures have changed very little with the inclusion of more women, still holding on to traditional masculine views of competition, and allowing the politicization and corporatization of the Games. It remains to be seen how the IOC will respond to these criticisms as well as to critiques of the binary categorization of athletes as yet another discriminatory pattern in the treatment of female athletes.

REFERENCES

Burstyn, V. (1998) 'The Politics of Globalization, Ideology, Gender and Olympic Sport'. *Global and Cultural Critique: Problematizing the Olympic Games*, 11–20, www.la84foundation.org/SportsLibrary/ISOR/ISOR1998e.pdf (accessed 14 December 2008).

Carpentier, F. and J.-P. Lefèvre (2006) 'The Modern Olympic Movement, Women's Sport and the Social Order During the Inter-war Period'. *The International Journal of the History of Sport* 23(7): 1,112–27.

Cavanagh, S. and H. Sykes (2006) 'Transsexual Bodies at the Olympics: The International Olympic Committee's Policy on Transsexual Athletes at the 2004 Athens Summer Games'. *Body & Society* 12(3): 75–102.

Chase, L. (1992) 'A Policy Analysis of Gender Inequality within the Olympic Movement'. *Proceedings: First International Symposium for Olympic Research*, 28–39, www.la84foundation.org/SportsLibrary/ISOR/ISOR1992g.pdf (accessed 1 December 2008).

DeFrantz, A. (1997) 'The Changing Role of Women in the Olympic Games'. *Olympic Review* 26(15): 18–21.

Dyer, K. (1982) *Challenging the Men: Women in Sport.* St Lucia: University of Queensland Press.

Eastman, S. and A. Billings (1999) 'Gender Parity in the Olympics: Hyping Women Athletes, Favoring Men Athletes'. *Journal of Sport and Social Issues* 23(2): 140–70.

Guttmann, A. (1991) *Women's Sports: A History.* New York: Columbia University Press.

Hargreaves, J. (1994) *Sporting Females: Critical Issues in the History and Sociology of Women's Sports.* London: Routledge.

Holder, D. (1996) 'A Woman's Place is at the Games'. *The Independent*, www.independ ent.co.uk/arts-entertainment/a-womans-place-is-at-the-games-1327620.html (accessed 3 January 2009).

IOC (2009) *Evolution of Women in Sport*, www.olympic.org/uk/organisation/missions/ women/evolution_uk.asp (accessed 1 March 2009).

Leigh, M. (1974) 'Pierre de Coubertin: A Man of His Time'. *Quest* 22(1): 19–24.

——(1980) 'The Enigma of Avery Brundage and Women Athletes'. *Arena Review* 4: 11–21.

Leigh, M. and M. Bonin (1974) 'The Pioneering Role of Madame Alice Milliat and the FSFI in Establishing International Trade [sic] and Field Competition for Women'. *Journal of Sport History* 4(1): 72–83.

Mitchell, S. (1977) 'Women's Participation in the Olympic Games 1900–1926'. *Journal of Sport History* 4(s2): 208–28.

Pittaway, B. (1999) 'Olympic Bosses Suspend Sex Tests'. *The Express*, www.pfc.org.uk/ node/857 (accessed 1 March 2009).

Samaranch, J. (1981) 'The IOC President Informs the Ifs'. *Olympic Review* 170: 694–97.

Schneider, A. (2000) 'Olympic Reform, Are We There Yet?' *Bridging Three Centuries: Intellectual Crossroads and the Modern Olympic Movement*, 225–32, www. la84foundation.org/SportsLibrary/ISOR/ISOR2000zb.pdf (accessed 12 January 2009).

Simri, U. (1979) *Women at the Olympics.* Netanya, Israel: The Wingate Institute for Physical Education and Sport.

Spears, B. (1976) 'Women in the Olympics: An Unresolved Problem', in P. Graham and H. Uberhorst (eds), *The Modern Olympics.* West Point, NY: Leisure Press, 53–73.

Sykes, H. (2006) 'Transsexual and Transgender Policies in Sport'. *Women in Sport and Physical Activity Journal* 15(1): 3–13.

Teetzel, S. (2006) 'Equality, Equity, and Inclusion: Issues in Women and Transgendered Athletes', in N. Crowter, M. Heine and R. K. Barney (eds), *Cultural Imperialism in Action: Critiques in the Global Olympic Trust*, 331–38, www. la84foundation.org/SportsLibrary/ISOR/ISOR2006ae.pdf (accessed 4 January 2009).

Wackwitz, L. (2003) 'Verifying the Myth: Olympic Sex Testing and the Category "Woman"'. *Women's Studies International Forum*, 26(6): 553–60.

Wilson, W. (1996) 'The IOC and the Status of Women in the Olympic Movement: 1972–96'. *Research Quarterly for Exercise and Sport* 67(2): 183–92.

Wushanley, Y. (2000) 'The Olympics, Cold War, and the Reconstruction of U.S. Women's Athletics'. *Bridging Three Centuries: Intellectual Crossroads and the Modern Olympic Movement*, 119–26, www.la84foundation.org/SportsLibrary/ISOR/ ISOR2000p.pdf (accessed 3 January 2009).

FURTHER READING

Leder, J. M. (1996) *Grace and Glory: Century of Women in the Olympics*. Chicago: Triumph Books.

Markula, P. (ed.) (2009) *Olympic Women and the Media: International Perspectives*. Basingstoke: Palgrave Macmillan.

Talbot, M. (ed.) (2001) *Gender, Power and Culture: Centenary Celebration of Women in the Olympics*. Oxford: Meyer & Meyer.

The politics of performance enhancement in the Olympic Games

ROB BEAMISH

INTRODUCTION

The positions in the politics of drug use in the Olympic Games seem to be crystal clear. The Honourable Charles Dubin, conducting a federal investigation into the use of drugs in sport following Ben Johnson's disqualification for a positive test at the 1988 Games opened his report with the following: 'The use of banned performance-enhancing drugs is cheating which is the antithesis of sport. The widespread use of such drugs', he continued, 'has threatened the essential integrity of sport and is destructive of its very objectives. It also erodes the ethical and moral values of athletes who use them, endangering their mental and physical welfare while demoralizing the entire sport community' (cited in Beamish and Ritchie, 2006: 111).

Some 15 years later, the World Anti-Doping Agency (WADA) supported those sentiments: 'Anti-doping programs seek to preserve what is intrinsically valuable about sport. The intrinsic value is often referred to as "the spirit of sport".' That spirit is characterized by 'ethics, fair play and honesty; health; excellence in performance; character and education; fun and joy; teamwork; dedication and commitment; respect for rules and laws; respect for self and other participants; courage; community and solidarity'. WADA continued: 'Doping is fundamentally contrary to the spirit of sport' (cited in Beamish and Ritchie, 2006: 111).

On the other hand, former International Olympic Committee (IOC) President Juan Antonio Samaranch offered the following assessment during a 'doping crisis' in the 1998 Tour de France. 'Doping is everything that, firstly, is harmful to an athlete's health and, secondly, artificially augments his performance.' However, he continued, 'if it's the second case, for me, it's not doping. If it's the first case, it is' (cited in Beamish and Ritchie, 2006: 1). More recently, Kayser and Smith (2008: 85), with the support of more than 30 other sport scientists, have argued that 'there are compelling reasons' to question WADA and UNESCO's current efforts in 'the globalisation and harmonisation of anti-doping'.

The division seems clear, but it is still somewhat murky. Using 'banned performance-enhancing substances' is cheating; if certain substances are against the rules then their use cheats other competitors staying within the rules. That much is not contentious.

58

The 'spirit of sport', as WADA defines it, involves ethics, fair play, honesty, character, respect for rules, and respect for self and other participants. It also involves health, excellence in performance, teamwork, dedication and commitment, and courage. So, if no substances were banned, would there be a problem? Would the use of performance-enhancing substances continue to be contrary to the spirit of sport defined by WADA? Would the two apparently opposing sides in the politics of 'drugs in sport' be brought together? The answer, unfortunately, is not clear either—it is more a 'yes and no' than a straightforward 'yes'.

SPORT, MODERNITY AND THE OLYMPIC GAMES

There are several detailed studies and synoptic overviews that chronicle the main 'events' in the use of drugs in the Olympic Games and the politics of the current ban on some performance-enhancing substances. This chapter will not retrace those events in detail, but focus instead on the fundamental basis for the Olympic Movement's original, principled ban of performance-enhancing practices, the deterioration and eventual abandoning of those principles, the ongoing rhetorical support for banning selected practices and substances based on those abandoned principles, and the reality within which the ban can be principally removed so that the 'spirit' of contemporary sport, as it is practised, may be openly and honestly honoured.

At the most fundamental level, the current regulation of performance-enhancing substances and practices is the legacy of Baron Pierre de Coubertin's desire to use the modern Olympic Games as one mechanism to halt the march of modernity—a quest that failed—versus the social forces advancing modernity and the extent to which instrumental reason (i.e. the calculation of the best means to an end) is central to the modernist project.

Modernity has its roots in the Renaissance and Enlightenment but began to visibly emerge in 19th-century Western Europe. Modernity is strongly associated with the dismissal or marginalization of tradition and a firm belief in and commitment to progress. Within modernity, there is an increasing reliance upon the systematic use of human reason and scientific knowledge to create technologies and social arrangements that will permit greater secular freedom and more precise control of the social and natural worlds. These latter elements constitute the most significant 'promise' of modernism—the potential for humankind to control its future development through the use of scientifically informed, instrumental reason.

The politics of banned performance-enhancing substances boils down, as the opening quotations indicate, to how one chooses to define 'the spirit of sport' and that depends upon how one answers the question, 'what is "sport" really?' Is 'sport' something that exists outside of history? Is it a form of human activity that possesses the same essential characteristics in every period of human history? Or is 'sport' a form of physical activity that is shaped and defined by the different periods of history in which people undertake it? The nature of 'sport' would then emerge from the way athletes produce it and the way sport administrators within that particular social period, try to regulate it. The context and tenor of a particular social period and the socially constituted meaning of sport would then define 'the spirit of sport'.

In the end, the political debate over performance-enhancing substances at the Games stems from de Coubertin's attempt to define 'sport' and 'the spirit of sport' in trans-historical terms—as entities that have the same essential characteristics across all of human history. The content of sport, de Coubertin argued, was defined by a particular, chivalric code, brothers-in-arms and the ethos of 'fair play'. It was this form of sport that de Coubertin sought to use to resist the march of modernity across Europe.

While de Coubertin and all ensuing IOC presidents failed in their attempts to use sport to overturn modernity, they did succeed in keeping key aspects of de Coubertin's rhetoric alive and it is this rhetoric that provides the basis for rejecting performance-enhancement in sport. As a result, the politics of performance-enhancing substances centre on an attempt to impose a moribund set of principles on a completely incompatible set of firmly established modernist sport practices. This is unfortunate because the politics of drugs in sport should address the genuinely pressing problems stemming from the reality of world-class sport in the modern period rather than 'tilting at windmills'.

SPORT: FROM ITS GENERIC SIMPLICITY TO A COMPLEX REALITY

At the most generic level, 'sport is an embodied, structured, goal-oriented, competitive, contest-based, ludic, physical activity' (Loy and Coakley, 2008). When people begin to play sport, they give it a reality (a form and substance). Thus 'sport' is shaped, executed and given meaning within a specific social context and is inescapably contoured by political forces that range from the formal—governments, sports governing bodies, education systems, etc.—to the informal practices of everyday life. Five important issues arise from these points.

First, any conception of 'the spirit of sport' must only be tied to the generic elements of sport as 'an embodied, structured, goal-oriented, competitive, contest-based, ludic, physical activity'. For de Coubertin and WADA, the 'spirit of sport' is tied to several additional qualities that they would like to establish and see in sport, but those qualities lie outside the essential qualities of sport. De Coubertin's (and WADA's) conception of the 'spirit of sport' involves politically motivated additions to what sport should be as opposed to what it was becoming and has become within modernity.

Second, despite any claim to the contrary, sport is politicized—a point made throughout this volume. The claim that sport or the Olympics are not political has always been a political claim in itself. 'Politics' is simply the formal and informal processes through which resources are distributed in a society. Sport could not exist without resources—it has always drawn from the social surplus. At the same time, sport contributes to the material, social and cultural wealth and richness of nations and communities. As sport expands on to the world stage, it becomes more involved in struggles over the resources upon which it may draw, those it must give (or return) to the social whole and those it may keep to consolidate or enhance its position within society. There is also a struggle over sport's symbolic meaning which impacts directly on the use (or ban) of performance-enhancing practices and substances.

Third, as an embodied, goal-oriented, competitive, physical activity, sport in the modern period has centred to an important extent on the politics of the human body. Practising a competitive, goal-oriented, physical activity has far-reaching consequences for athletes' bodies. Striving to be on the top of the world-class, high-performance sport pyramid demands physical (embodied) performances at the outer limits of human achievement. This requires demanding training regimes, increasingly sophisticated techniques and regimens of recovery, and the use of the best performance-enhancing practices possible. It is these demands that Samaranch addressed directly as do Kayser and Smith (2008) and others.

Faced with the realities of the ongoing pursuit of athletic performance at the outer limits of human achievement, formal sports governing bodies have faced two conflicting demands. On the one hand, some sports bodies must marshal resources that will assist the pursuit of athletic performance at the highest level. On the other, some sports governing bodies have the responsibility of attempting—or creating the impression that they are attempting—to regulate the real costs and risks of competitive, goal-oriented sport on athletes' bodies. Frequently—and certainly in the case of the IOC, most National Olympic Committees (NOCs) and numerous national sports governing bodies—the same formal organization must pursue, or appear to be pursuing, both of these conflicting objectives simultaneously.

While many formal sports governing bodies are torn in two directions, the demands of sport practice are quite singular for the athlete. Competitive sport at the Olympic level virtually compels athletes to drive their bodies to ever higher levels of performance within the regulated structure of sport. At the same time, athletes are pressured to go beyond the regulated parameters or move into uncharted territory in the quest for victory. More important, as the direct producers of sport, athletes develop their own culturally based sense of what is legitimate—what is consistent with their craft and what they would maintain is within 'the real spirit' of the high-performance sport world they produce.

Due to these different tensions and contradictions, there is an inescapable conflict between the intrinsic, logical requirements for success in sport as an 'embodied, structured, goal-oriented, competitive, contest-based' activity in the modern period, and the formal regulation of those practices—a regulation that must facilitate the intrinsic logic of competitive performance but also restrain it. Sport practices in the modern era inescapably lead to questions of performance-enhancing practices and those, in turn, to questions about the use and regulation of exogenous substances such as drugs in the pursuit of athletic victory. Determining what risks arise from the intrinsic logic of competitive sport at a world-class level, how they are best addressed and what those questions mean for the use of performance-enhancing substances by high-performance athletes are the most pressing questions in the politics of drug use in world-class, high-performance sport. Those should be the focal point of attention but too often they are overshadowed by 'de Coubertin's legacy'.

Fourth, the actual practice of any generic activity is never singular. The form and substance of 'sport' within the modern era varies significantly although some specific forms have come to dominate as they have captured more resources at the cost of other activities. In the modern period, the highly demanding, heavily resourced form of world-class, high-performance sport practised at the Olympics has become one of the most dominant forms of sport around the globe. This has had political consequences for 'sport', the Olympics, world-class, high-performance athletes, and the formal regulators of sport.

Because the Olympics are a spectacle in which the world's best athletes compete at the outer limits of athletic achievement, the Games have increasingly drawn from and ultimately relied upon specialized sport sciences, sophisticated technology and elaborate, heavily resourced systems of 'athlete production'. Performance-enhancing substances—some accepted and others banned—are one small element within that mixture. The formal and informal political pressures to use exogenous substances have grown as the Games have become increasingly spectacular (in every sense of the word).

Finally, although sport is an embodied physical activity, its two greatest values in the modern era are its symbolic power and its economic value. The modern Olympics began as a project in which symbolism was paramount— one in which, ironically, the Games were to demonstrate the strength and superiority of traditional values over the growing significance of modernity's scientific rationality and the reduction of human relationships to 'the cash nexus' of the market economy where everything, including the relationships between individuals, was perceived in terms of its economic (or cash) value.

The Games were also aimed at eliminating narrow forms of chauvinistic nationalism by promoting international harmony. In fact, rather than returning the globe to traditional values and eliminating chauvinist nationalism, from 1936 onwards the Olympics have heightened nationalism and expanded the commercialization of sport. The symbolic and modernist dimensions of the Games are central to their increased politicization and the use of performance-enhancing substances and practices. Both of these points merit some further exploration.

DE COUBERTIN AND THE FIGHT AGAINST MODERNITY

Attempts to ban drug use in the Olympic Games stem from the 'legacy of intent' that de Coubertin left to the movement. De Coubertin wanted to halt the unfolding reality of modernity—no modest task. Not everyone has embraced modernity and various influential 19th-century thinkers sought to stem the torrent of change and return Europe to its established, traditional values. However, de Coubertin was one of modernity's most strident opponents and no aspect of his Olympic project, its inspiration, its successes and its failures, can be understood without situating it within this context.

Deeply opposed to the growing materialism of industrial capitalism and its divisive competitive forces, de Coubertin wanted to forge a moral elite that would revitalize Europe and guide it back to the great cultural values of Western civilization. De Coubertin found much of his inspiration in the 'muscular Christian' philosophies of Canon Kingsley and Thomas Arnold, along with the religious spiritualism of the Ancient Games. Believing that character was not formed solely by the mind but also involved the body, de Coubertin launched the modern Games and attempted to control the embodied forms of sport practice within them so that the experience would create the specific, delicate balance of mind and body he valued so highly.

Through the Games, de Coubertin wanted to create 'brothers-in-arms' whose character would be forged in the 'struggle' sport provided. The traditional conceptions of chivalry, rather than the modernist emphasis upon victory, would serve as the code of conduct—the 'spirit of sport'. The calculated pursuit of victory was anathema to the spirit of sport as de Coubertin saw it.

The true amateur sportsman of the 19th-century United Kingdom served as the prototype for the athletes of the Games. Through the Olympics, de Coubertin and the IOC would transport the principles of 'playing on' from the United Kingdom's sports fields to the rest of the world, stem the tide of modernity and return Europe to the classical traditions of Western civilization. This is the sport form that de Coubertin, long before WADA and UNESCO, sought to harmonize around the globe. Sport as a character building, chivalric contest was de Coubertin's dream; it was his political project. This particular perception of sport continues to serve as the central reference point in the debates over performance-enhancing substances.

Despite his lofty goals, de Coubertin had to make concessions to the modern world he wanted to change. In contrast to his desire to combat the crass materialism of the market, early Olympiads—Paris, St Louis and London in 1900, 1904 and 1908, respectively—took place within the context of international exhibitions celebrating technology, science, industrial capitalism and modern culture more generally. The organizers of the 1928 Amsterdam Games sold rights packages to various companies, including Coca-Cola, bringing commercialism directly into the Olympic venues.

De Coubertin struggled with the amateur criterion, too. Football (soccer) was originally excluded from the Games because it was so thoroughly professionalized. Even the modest proposal that players should receive payments for lost wages was rejected because the powerful International Association of Athletics Federation threatened to abandon the Games if the IOC let the 'British standard' for amateurism slip in the slightest. At the same time, different nations and their NOCs saw the Games as the perfect opportunity to demonstrate their strength and vitality leading to the use of athletes who did not meet the Games' strict criteria of amateur. The 'amateur question' would plague the Games until 1974 when the IOC dropped the restrictive criteria through changes to Rule 26 (now Rule 41) of the *Olympic Charter*. After 1974 the Games were increasingly opened to the best athletes in the world— professionalized 'amateurs' and the overtly professional alike.

Prior to 1936 the Olympics were a marginal sporting event, but the Berlin Games changed their status dramatically. The Nazi Party had routinely exploited the new technologies of the mass media in spreading their propaganda domestically. They used imposing, emotion-laden, Wagner-inspired *Gesamtkunstwerke*—total works of art—in huge venues and sports stadiums across Germany to spread their ideology. Music, choreography, drama and neoclassical architecture were blended into captivating, exhilarating and emotionally draining experiences. The Nazis used the 1936 Games to weld the Promethean symbolic power of the Olympics to an image of raw power that loomed, like a spectre, over Europe.

The 1942 Games were cancelled because of the Second World War and, while the 1948 London Games inspired hope, they laboured under the burden of post-War recovery. The 1950s, however, opened up a quarter century of economic expansion, increased prosperity within the developed world, and began fulfilling the 'promise of modernity'. The increasing expansion of the market to all parts of social life was virtually unchallenged as Western Europeans and North Americans entered into the 'consumer's republic'. Unable to resist the forces of modernity, the Games were slowly overwhelmed by its social forces. Three interrelated developments from the 1950s onwards shaped Olympic sport—the increased professionalization of the Games' athletes, the Cold War turn towards heavily resourced, national, high-performance sports systems, and the increased use of science and technology in athlete preparation.

In the immediate post-Second World War period, international tensions between the West's liberal democracies and the state socialist societies of the East escalated rapidly. In the face of a 'no-holds-barred' athletic confrontation between the USA and the USSR, IOC President Avery Brundage struggled to keep de Coubertin's principles alive. Throughout his presidency, Brundage continually proclaimed that the Games were the best vehicle for advancing a morality of 'fair play', the 'spirit of rules kept' and internationalism. What catapulted the Games to prominence, however, was not their moral potential but the political drama of the East–West confrontation, gold medal counts and the entertainment spectacle of Cold War athletes engaged in the all-out pursuit of victory. As the stakes in the Cold War Games rose, nations were reluctant to leave their fate in the hands of part-time amateurs.

The IOC fought the trend, introducing Rule 26, 'The Amateurism Code', to its *Charter* in 1962, but that did not stop modernity's intrusion on both sides of the Iron Curtain. Training regimes throughout the 1960s became more demanding, scientifically programmed and closely monitored. As NOCs saw their athletes fall behind, they copied the high performance sport systems that had originated in the Eastern Bloc and began spreading throughout the West.

Faced with the rapid transformation of athlete preparation, a joint IOC/ NOC commission reviewed the 'eligibility question' in 1969 and 1970. The IOC changed Rule 26 from 'The Amateur Code' to 'The Eligibility Code' but kept the rule as restrictive as ever. In 1974, however, the IOC chose to abandon the movement's cardinal principle and eased the criteria for Rule 26. The new rule did more than fundamentally change the type of athletes participating in the Games; it opened the way to athletes whose motivations, commitment and actions were directly antithetical to Coubertin's 'spirit of sport'.

With one decision, the IOC had overturned the fundamental reason for reviving the ancient Olympic Games and opened its arms to commercial interests, an unrestricted competitive zeal and fully professionalized athletes. Post-1974, the Games would legitimately feature athletes for whom sport was a full-time, year-round commitment. Winning and the conquest of the linear record became the real 'spirit of sport' for the Games and the scientifically rational, technologically assisted pursuit of the limits to human physical performance became their central ethos. The 1974 decision legitimated open access to the financial and other resources needed for the pursuit of Olympic gold. Despite the 1974 decision to open the Games to fully professional and professionalized high-performance athletes, the IOC maintained its rhetorical claim to the purity of sport while simultaneously thriving on the full modernization of the athletic spectacle it organized and fully controlled.

WINNING IN THE COLD WAR GAMES

Moving in concert with the professionalization of Olympic athletes was the spread of high-performance sport systems. Joseph Stalin and the USSR were the first to systematically develop 'Cold War athlete-warriors', but Walter Ulbricht in the German Democratic Republic (GDR, or East Germany) was quick to follow. Whereas the USSR tended to rely on the size of its population to produce enough athletes for the system to develop world champions, the GDR made more strategic investments and quickly developed one of the most heavily resourced and sophisticated systems of applied 'athlete production'.

The Eastern Bloc was not alone. When the IOC granted the 1972 Games to the Federal Republic of Germany (FRG, or West Germany), the FRG began pouring resources into its own national sport system. Canada would follow suit when awarded the 1976 Games, borrowing aspects from both German systems and moulding them to Canada's own unique federal structure. Australia would build on and improve the Canadian model. In each instance, national governments invested heavily in an athlete development infrastructure with the primary objective of enhancing performance at the Olympic Games. China's vast investment in every aspect of the systematic production of world-class, medal winning athletes for 2008 was simply the most recent example of a national sport system although its scale positively dwarfed all those that preceded it.

Drug use in sport extends well back into the pre-Second World War period but it was during the 1950s and 1960s that individuals with specific Cold War sport interests began to pursue seriously the use of pharmaceuticals that would significantly enhance performance. While American team physician John Ziegler has developed an elaborate story about the first Soviet use of steroids and his own ensuing development of Dianabol with Ciba Pharmaceutical Company, Dimeo (2007: 72–77) has argued that steroid use did not simply begin with the USSR in 1952 to be duplicated by the Americans shortly thereafter to level the playing field.

The use of steroids in high-performance sport was part of a process that involved refining the synthetic hormone, convincing athletes that synthetic steroids would increase power and to use them, and developing training programmes that would produce significant results. The early results were mixed but within the high stakes of East–West competition of the 1950s physicians, pharmaceutical firms, coaches and other interested parties were able to provide athletes with steroids and training regimes that significantly enhanced performance in the power sports. Their use quickly spread and by the mid-1960s, steroids were almost commonplace.

While steroids are now the main focus in the politics of drug use at the Games, it was the death of Danish cyclist Knud Jensen at the 1960 Summer Games that forced the IOC to seriously address the question of performance-enhancing substances. The IOC quickly established its first Medical Committee to investigate drug use in sport.

The Committee used de Coubertin's 'spirit of sport' as its reference point despite the total and complete infusion of modernist practices into world-class, high-performance sport. As a result, the cautious, conservative medical experts on the Committee recommended that the IOC condemn drug use, introduce athlete testing, have athletes sign a pledge of non-use, and sanction individuals and NSOs implicated in drug use (see Dimeo, 2007). In 1967 the IOC prohibited 'the use of substances or techniques in any form or quantity alien or unnatural to the body with the exclusive aim of obtaining an artificial or unfair increase of performance in competition' and drafted a list of banned substances that ranged from cocaine, pep pills and vasodilators, to alcohol, opiates and hashish (Todd and Todd, 2001: 68). Steroids were included in a note to the list. Despite the ban, the IOC did not have a suitable test for steroids until 1973 and did not test for them until the 1976 Games.

In the interval between 1952 and 1976, world-class sport had become the all-out assault on the linear record and the ban had nothing more than the moral authority of the empty rhetoric of de Coubertin's long since bypassed 'spirit of sport'. In the first quarter century of the post-Second World War period, the Games had become a fully open competition among state-supported athlete-warriors who drew from the most advanced scientific knowledge to increase their chances of Olympic gold. During this period, athletes and the architects of national sport systems had increasingly embraced all that modernity had to offer including an overwhelming zeal for athletic victory supported by cutting-edge, scientific knowledge. The real framing for the 'politics of performance-enhancing substances' was in place by 1976 but it was obscured by the 'legacy of intent' within de Coubertin's particular 'spirit of sport'. That legacy, in conflict with the actual modernist 'spirit of sport' that was firmly established by the 1970s, has stood as the political lightning rod in the debate over performance-enhancing substances and practices.

THE POLITICAL CROSSROADS OF PERFORMANCE ENHANCEMENT

From 1976 onwards, the politics of performance-enhancing substances has been unable to escape the dead hand of de Coubertin's aspirations and legacy of intent and this has come at a tremendous cost to high-performance athletes as a group, specific individual athletes in particular, and the credibility of the Games and the IOC. The politics of performance enhancement really centre on two specific questions. First, does the IOC, or the world sport community, want to renew the pursuit of de Coubertin's project and its particular, very specific, 'spirit of sport'? If that is the case, then the world sport community will have to struggle against the entrenched realities of modernity and re-launch the de Coubertin project in all of its purity of aspiration. There are certainly anti-modernist forces in various parts of the globe and anti-modernists in certain niches of the broad spectrum of sport forms in the contemporary period, but the task would be monumental and has already failed once. So the question arises: which image of sport does the IOC want to support in the Games—de Coubertin's traditionalist, chivalric 'spirit of sport', or the now-dominant modernist spirit?

The second question follows directly from the first: in what manner should performance-enhancing practices be regulated within the sport form chosen? The list of banned performance-enhancing substances and practices will necessarily grow if the IOC or the world sport community opt for a renewed attempt at establishing de Coubertin's project. Opting for the modernist reality would pose very different problems—a very different political dynamic.

The generic nature of sport, manifested within modernity, encourages, facilitates and almost compels athletes to pursue athletic achievement at the outer limits of human capacity. It is heavily resourced and supported by national sport systems, advanced technology and the latest scientific research. As the limits of performance are pushed closer to the absolute edge of possibility, however, the risk of harm to the athlete grows exponentially. World-class athletes have no choice but to pursue the logic of performance. Sports administrators must balance between facilitation and the regulation of the costs and risks that competitive, goal-oriented sport has upon athletes' bodies. Here the politics of performance-enhancing substances and practices involves the development of the best harm-reduction strategies possible; strategies and policies that remain consistent with the modernist ethos of world-class, high-performance sport but place athletes' health and safety ahead of victory. This debate should not be put off any longer.

REFERENCES

Beamish, R. and I. Ritchie (2006) *Fastest, Highest, Strongest: A Critique of High-Performance Sport*. London: Routledge.

Coubertin, P. (2000) *Olympism: Selected Writings*. Lausanne: International Olympic Committee.

Dimeo, P. (2007) *A History of Drug Use in Sport: 1876–1976*. London: Routledge.

Dubin, C. (1990) *Commission of Inquiry into the Use of Drugs and Banned Practices Intended to Increase Athletic Performance*. Ottawa: Canadian Government Publishing Centre.

Hoberman, J. (2006) *Testosterone Dreams: Rejuvenation, Aphrodisia, Doping*. Berkeley, CA: University of California Press.

Kayser, B. and A. Smith (2008) 'Globalisation of Anti-doping: The Reverse Side of the Medal'. *The British Medical Journal* 337 (July): 85–87.

Loy, J.W. and J. Coakley (2008) 'Sport', in G. Ritzer (ed.), *Blackwell Encyclopedia of Sociology*. Blackwell Publishing. Blackwell Reference Online, www.black-wellreference.com/subscriber/tocnode?id=g9781405124331_chunk_g978140512433125_ss1–221 (accessed 6 September 2008).

Møller, V. (2008) *The Doping Devil*. Copenhagen: International Network of Humanistic Doping Research.

Todd, J. and T. Todd (2001) 'Significant Events in the History of Drug Testing and the Olympic Movement: 1960–99', in W. Wilson and E. Derse (eds), *Doping in Elite Sport: The Politics of Drugs in the Olympic Movement*. Champaign, IL: Human Kinetics Press, 65–128.

Voy, R. (1991) *Drugs, Sport and Politics*. Champaign, IL: Leisure Press.

Wagg, S. and D. Andrews (eds) (2007) *East Plays West: Sport and the Cold War*. London: Routledge.

Yesalis, C. and M. Bahrke (2002) 'History of Doping in Sport'. *International Sports Studies* 24(1): 42–76.

FURTHER READING

Burns, C.N. (2005) *Doping in Sport*. Hauppauge, NY: Nova Science Publishers.

Hatton, C. (2008) *The Night Olympic Team: Fighting to Keep Drugs Out of the Games*. Homesdale, PA: Boyds Mills Press.

Wilson, W. (2000) *Doping in Elite Sport: The Politics of Drugs and the Olympic Movement*. Champaign, IL: Human Kinetics.

Disability, Olympism and Paralympism

P. David Howe

INTRODUCTION

People with physical and sensory disabilities have been competing in international sports for over 80 years. The public is still often led to believe that sport events for disabled people are so virtuous that they are devoid of politics and beyond petty political argument because they are an all-inclusive, wholesome invention. This myth associated with sport for the disabled should be critically examined as politics are fundamental to all spheres of society and, consequently, have been central to sport for the disabled from its genesis.

Events for athletes with hearing disability were organized at an international level as early as 1924 (Doll-Tepper, 1999). An event, known as the International Silent Games, was held in Paris and became the first staging of what are now called the Deaflympics. The first international competition for those who use wheelchairs was staged in 1952 at Stoke Mandeville, England. These early events were staged for disability groups who 'suffered' from a single impairment and were established with a charitable mandate to help a particular marginalized group of people. The fact that early events were disability and not sport specific led to ever increasing political tension between various International Organizations of Sport for the Disabled (IOSDs) prior to the establishment of the International Paralympic Committee (IPC) in 1989. The IOSDs, including the Cerebral Palsy International Sport and Recreation Association (CP-ISRA), the International Blind Sport Association (IBSA), the International Sports Federation for Persons with Intellectual Disability (INAS-FID), and the International Wheelchair and Amputee Sport Association (IWAS), were instrumental in the development of the IPC. IPC events have been open to individual members of the IOSDs but not all disabilities are catered for within these groups and, thus, many forms of disability are excluded from these events. The politics surrounding which disability groups are part of the Paralympic Movement and how much influence each group has within it is central to this discussion.

This chapter explores a number of key developments within disability sport generally and Paralympic sport in particular that have impacted upon the cultural politics of the Paralympic Movement (see Howe, 2008). This critical evaluation starts by discussing the importance of the political agenda behind the development of the academic field of disability studies which has led to a lack of attention being paid to the embodied experience of disability. By highlighting the changes within sport for the disabled governance, the chapter will also draw attention to the debate surrounding the use of Olympic values, articulated through the ideology of Olympism, as a way of possibly adding virtue to the Paralympic Movement. The ideology of Paralympism fostered by the IPC is related to, but distinct from, Olympism because of the centrality of the praxis of categorizing disabled bodies known as classification. Finally, classification will be shown to be a political as well as a mechanical process for ordering disabled bodies before competition even begins.

DISABILITY STUDIES: THE DELIBERATE AVOIDANCE OF SPORT

The development of the interdisciplinary field of disability studies is a by-product of political activism within the disabled community that attempted to establish 'the disabled' as a homogeneous collective who are oppressed by able society (Zola, 1981). In his classic text, *The Politics of Disablement,* Michael Oliver (1990) suggests that a disability has been defined traditionally as an individual loss or restriction in terms of ability and claims that this is politically problematic. Following Oliver (1990), social scientists working in the field of disability studies began to criticize the way in which society had traditionally treated individuals with disabilities. In particular, these scholars believe that the categorization of disabled people using a medical system leads to them being singled out as different from normal healthy people and, ultimately, as inferior. Here the link between political activism and the development of an academic discipline is clear (Campbell and Oliver, 1996). Within the literature, the process of categorizing impaired bodies was and is referred to as the medical model of disability. Disability, in other words, is seen as a medical problem because people with disabilities have traditionally been treated as if they are permanently ill.

In order to improve the lives of the disabled, activists who became the founding figures within disability studies began to argue that being disabled was not like being ill but was the result of society constructing this group in this manner. This led to the development of the 'social model of disability' which suggests that any negative impact on a disabled person's life that is a result of their disability (i.e. the physical or sensory lack of bodily function a person may have) is a direct consequence of how society is structured. Those who are involved politically in disability advocacy as a result of the development of the social model have celebrated the work of neo-Marxist thinkers like Oliver (1990) by doing two things: articulating the disabled community as homogeneous and eliminating the body and, thus, specific disabilities from the debate. The main argument is that because it is the body that marginalizes the disabled community, it should be removed from the political debate (Campbell and Oliver, 1996). The elimination of the body is problematic as disability per se brings the physical body into focus, but also because the embryonic stage of development of the Paralympic Games was rehabilitative sport.

Many, if not all, individuals who acquire a disability through an accident and the vast majority of those who are born with congenital deformity are enrolled in rehabilitative programmes that are designed to minimize the influence of the abnormal body. In such programmes, sport is seen in a functional way as instrumental to a disabled person's socialization. In addition, sport for the disabled is organized through a process known as classification that is biomedical in origin, though political in practice. It is perhaps not surprising then that disability activists and scholars have distanced themselves from critical examination of sport for the disabled because to them the problems with such institutions are self-evident. Yet within the social scientific research into sport there have been calls for a closer and critical examination of the political environment surrounding Paralympic sport (Howe, 2008). This examination begins with an overview of the governance of Paralympic sports.

THE GOVERNANCE OF PARALYMPIC SPORTS

The Paralympic Movement is generally considered to have begun life as a direct result of the use of sport in the years following the Second World War as a form of rehabilitation of men that had been impaired as a result of participation in military activity (Scruton, 1998). A key moment in the development of disability sport took place in 1948 at Stoke Mandeville Hospital in England, where a doctor, Sir Ludwig Guttmann, organized a sports event for his patients who were all wheelchair users. Coinciding with the opening ceremonies of that year's Olympic Games in London, the event included sports such as archery and field events and is widely regarded as the early ancestor of the modern Paralympic Games, which is now regarded as the most important event on the international sporting calendar for people with different types of disability.

The legacy of the event run at Stoke Mandeville hospital can be linked directly to the international federation responsible for the organization of events for sports people in wheelchairs. The International Stoke Mandeville Games Wheelchair Sports Federation (ISMWSF) organized many of the key events in a variety of sports played throughout the world for well over half a century. Since 2004, this organization has been called International Wheelchair and Amputee Sport (IWAS), with a remit for both athletes in wheelchairs and amputees. ISMWSF was and IWAS is amongst those IOSDs that independently developed systems of classification which are in essence a method of equitably categorizing bodies for the practice of competitive sport. The IOSDs were established with the explicit intention of creating opportunities for disabled people to be involved in the practice of sport as a vehicle for their empowerment (Howe, 2008). As a result, there has been considerable political infighting amongst the IOSDs to secure the best competitive opportunities for their athletes, sometimes at the expense of the other federations and the athletes they represent.

The IOSDs—the key objective of which was to facilitate participation in sport by their constituent disability groups—helped to organize the Paralympic Games from 1960 through to 1988 and, as a result, these games were different in that there was less emphasis on high performance even though some elite athletes were involved. Nevertheless, a hallmark of these games was the continual political infighting between the various IOSDs while each attempted to establish its authority over the management and structuring of the Paralympic Games. Before the establishment of the IPC, an organization known as the International Coordinating Committee (ICC) existed in 1982–89 with the aim of getting the IOSDs to communicate. The ICC was chaired by the president of one of the IOSDs for a period of six months at which point the position was handed on to another IOSD president. While this may seem an egalitarian way to run an international organization, the quick turnaround meant that little was achieved under this umbrella organization in terms of moving sport for the disabled forward.

The political turmoil that characterized early Paralympic Games led to the establishment of the IPC as a supposedly neutral organization that would act as the custodian of the games and champion of all the IOSDs. Establishing the IPC did not, however, end the conflict between the IOSDs. The IOSDs continue to position themselves so that they might retain some power and influence within the relatively high-profile world of Paralympic Sport. It was the ISMWSF which effectively gained control of the IPC in 1989 as the first president, Dr Robert Steadward, had been a leading figure in this organization. Early IPC concerns about a manageable classification system can be traced back to the fact that ISMWSF, the oldest IOSD, had the largest number of athletes and the streamlining of the classification system would not hinder this group's involvement in IPC events to the same extent that it would that of other IOSDs (Jones and Howe, 2005). The merger of the ISMWSF with the ISOD in 2004 (IWAS), has allowed this group to consolidate its influence within the IPC (see Howe, 2008). Not only does it control the greatest number of athletes within the IPC but the disabilities involved are relatively well understood as 'unfortunate' by the public at large making these athletes ideal role models for the Paralympic Movement.

The IPC currently organizes and administers both the Paralympic Games and the quadrennial world championships for individual Paralympic sports, such as swimming and athletics. Using the resources of the IOSDs (including athletes, volunteer administrators and classification systems), the IPC has arguably turned the Paralympic Games into the most recognizable and possibly most influential vehicle for the promotion of disability sport. The Paralympic Games are well organized with a relatively high profile that attracts significant media coverage and commercial sponsorship like many other modern sporting spectacles. Athletes from 136 nations competed in the 2004 Paralympic Games in Athens, making the Paralympic Games unquestionably the main international sporting forum for athletes with different disabilities.

However, this increased exposure has not been without its problems. The Paralympic Games now cater for a less diverse impaired population than it did in 1988 (Howe, 2008). It seems that the need for commercial success has had a detrimental impact on the most severely disabled competitors whose sporting opportunities have been steadily reduced as the Games have become a relatively large media spectacle. As the number of severely disabled athletes involved in the Paralympics has declined, so has the political influence of CP-ISRA, many of whose athletes no longer have events at the Paralympics for which political influence is determined in part by how many athletes each IOSD controls. While currently the IPC attracts attention for organizing a media spectacle, in the past it was viewed by the public as virtuous and closely linked to the ideals of Olympism. Although the Olympic ideology forms the cornerstone of Paralympism, within the IPC this is rooted in a desire to maintain a discrete political identity. Once it became clear that the Olympic Movement would not fully embrace the Paralympics (Labanowich, 1988), there emerged a need to justify sport for the disabled as a disability business. What better way to do this than establish an ideological foundation for the Paralympic Movement.

FROM OLYMPISM TO PARALYMPISM

The fundamental principles of the Olympic Movement enshrined in the *Olympic Charter*s (International Olympic Committee, 2007: 11) suggest that:

> Olympism is a philosophy of life, exalting and combining in a balanced whole the qualities of body, will and mind. Blending sport with culture and education, Olympism seeks to create a way of life based on the joy of effort, the educational value of good example and respect for universal fundamental ethical principles.

It is further suggested that:

> The goal of Olympism is to place sport at the service of the harmonious development of man, with a view to promoting a peaceful society concerned with the preservation of human dignity.

These virtues may be seen to be part of the Paralympic Movement at least by those with influence in the IOC. Over 50 years ago, during the 1956 Olympic Games, 'the International Olympic Committee awarded the Fearnley Cup to the organisers of the International Stoke Mandeville Games [the antecedent of the Paralympics] for "outstanding achievement in the service of Olympic ideals"' (Goodman, 1986: 157). Four years later, in 1960, at the closing ceremony of the first Paralympic Games in Rome, Guttmann was described by Pope John XXIII as the 'de Coubertin of the Paralyzed' (Scruton, 1998: xiii). Therefore, at the highest level, within the IOC, statements were made that publically supported the relatively new Paralympic Movement as a comrade that shared the same virtues as the Olympic Games.

However, since the Paralympic Movement has a distinctive cultural history, there is a need to establish an understanding of the virtues of Paralympism that are similar to but distinct from those of Olympism. There are those who advocate simply the use of Olympism as a sign of the growing harmony between the Olympic and Paralympic Movements because discrimination towards individuals is contrary to the principle tenets of Olympism (Labano-wich, 1988). In practice, however, both the Olympic and Paralympic Games exclude those who lack ability or, to put it another way, 'those who cannot'. While Olympism may be an inclusive ideology, the practice of high-performance sport is not.

The Paralympic Movement has raised public consciousness, both trans-nationally and trans-culturally, with respect to the philosophical concept and meaning of a *human* performance (Jones and Howe, 2005). It has allowed for the debate regarding Western sporting practice (ethos) and attributes virtue to those who achieve excellence in measurable sporting terms. In this respect, the achievements of Paralympians will never be of the same status as those of Olympians. While some commentators argued that the Paralympic Games should become part of the Olympics, political agreement between the IPC and the IOC has left the two events as separate if now more closely related entities that are legally bound to be marketed and promoted as a single product (IPC, 2003). If the two had become one there may have been a good argument for using Olympism as the ideological foundation for both movements.

One area where the Olympic and Paralympic Movements are distinct is entry into the competitive arena. In 'The Fundamentals of the Philosophy of the Modern Olympics', de Coubertin highlights the degree to which physical culture may be used as a vehicle to achieve sporting excellence when he states that 'For a hundred men to take part in physical education, you must have fifty who go in for sport. For fifty to go in for sport, you must have twenty to specialise, you must have five who are capable of remarkable physical feats' (de Coubertin 1956: 53). This is not a situation that is commonplace in Paralympic sport. In certain events at the Paralympic Games, such as track and field, selection is arguably tougher within class T54 (the most 'able' of wheelchair racers) than for the Olympic Games, in part because hundredths of seconds can separate those who win at the Paralympic Games and those who are not good enough to attend them. The closeness of races is a product of the technology used in developing racing wheelchairs but this does make for contests as spectacular as anything at the Olympics (Jones and Howe, 2005). However, outside of this particular class, it is relatively easier to be selected for the Paralympics than it is for the Olympic Games since fewer competitors are vying for the honour.

The Paralympic Games also cannot follow de Coubertin's vision of the cycle of Olympiad to provide the youth of any moment in time with the opportu-nity to compete in an international context. By stating that the 'Springtime of human life is found in the young adult who may be compared to a superb machine up and ready to enter, into full activity' (de Coubertin 1956: 53–54), de Coubertin demonstrated his belief that the Olympics were an ideal environment for fostering youth. Many athletes who compete in the Paralympic Games are 'eligible' as a result of a traumatic occurrence in life which clearly has no fixed time in the development of the individual. The active rehabilitation of people following injury where sport plays a key role can be seen as instrumental in creating another individual; almost a re-birth. In other words, Paralympians are generally older than Olympians.

With ongoing political conflicts around the globe in some cases resulting in wars over both ideology and geography, there would appear to be an increasing supply of causalities who need to be rehabilitated. Advances in medicine also mean that more soldiers are surviving after suffering trauma that would have been fatal at the time of the Second World War. The outcome of rehabilitative sport is not simply a route back into employment as had been Guttmann's aim (Anderson, 2003). Now disabled sportsmen and women do more than participate in high performance sport. They can now make a living at it in the process (Howe, 2008).

Though many governments around the world are increasing the funding that disability sports organizations receive, the structural opportunities for Paralympic athletes are still poor in comparison with those afforded to the Olympian. This can be manifested in what has been called the 'super-cripisation' of sport (see Howe, 2008: 106–7). Key to this understanding is that athletes involved within the Paralympic Games who are successful have become so in spite of their disability. Heart-warming stories of triumph over adversity are the calling cards of 'supercripping' the Paralympic Movement. Media coverage of recent Paralympic Games has played tribute to this phenomenon according to which heroic stories of survival dominate column inches set aside for the Games. Coverage in the media is seldom about the ability of sportspeople with disabilities, but rather about how the able bodied world can take inspiration from Paralympians simply because they are competing.

Within such celebrated achievements, key elements of Olympism are evident. The balancing of the body with a will can be seen to be played out in 'supercrip' fables. These stories celebrate the will of athletes that allows them to live as normal lives as possible. It has even become commonplace to see the classification of bodies into equitable categories as a normative activity. Such a process is of political importance since the IPC, though distancing itself from explicit discussion of Paralympism, has tried to make the process of classification appear as scientific and, therefore, as neutral as possible. The IPC adopts the dictum *Empower, Inspire, Achieve* and the vision, 'To Enable Paralympic Athletes to Achieve Sporting Excellence and Inspire and Excite the World', which are clearly important in establishing an ideology for the Paralympic Movement. However, at the heart of Paralympism is the system and process of classification that make the Paralympic Movement distinctive from mainstream sport and the ideals associated with the Olympic Games.

75

CLASSIFICATION: AT THE HEART OF PARALYMPISM

Classification is simply a structure, biomedical in nature, for competition similar to the systems used in the sports of judo and boxing where competitors perform in distinctive weight categories. Within disability sports, competitors are classified by the body's degree of function and it is, therefore, important that the classification process is robust and achieves equity across the Paralympic sporting practice and enables athletes to compete on a 'level playing field'. Establishing this level playing field is, however, highly political as there is never complete agreement among and within the IOSDs as to how to fairly and equitably classify the athletes.

The process of classification within sport for the disabled makes distinctions between the physical or mental potential of athletes and attempts to achieve an equitable environment whereby after competition the successful athletes in each class will have an equal chance of accumulating physical capital. In reality, however, a number of factors impact upon the accumulation of capital (both physical and cultural) in various classifications. The first factor is the number of athletes within a particular event. If there are only a handful of athletes, then the amount of capital that can be accumulated in most cases is limited. To the outside world the achievements of athletes in a highly contested event are generally seen as superior to those in events with a small number of competitors. In the case of the Paralympic Games, events for the severely disabled are under-subscribed and, as such, competing athletes are seen to lack capital. Indeed, it is these severely disabled athletes' events that have been removed from the Paralympic programme in recent years. Yet if Paralympism is to have a lasting virtue, it needs to consider the rights of the most disabled populations to experience international competition.

In some classes there may only be six athletes from four countries (the IPC minimum for eligible events), suggesting that winners are less likely to receive the same kudos as an athlete who defeats 20 competitors. Another important factor in terms of whether winners ultimately gain capital from their involvement in sport is the nature and degree of their disability. Some disabilities are more acceptable than others in part because they adhere more closely to the cultural norms of aesthetics that are associated with the able bodied. The relative proximity to the mainstream norm can also be linked to how the disabilities occurred. For example, there seems to be greater acceptance of acquired disabilities as opposed to those that are the result of congenital deformity.

There may be numerous cultural reasons for this hierarchy of acceptance for particular disabilities within mainstream society. It might be easier to relate to someone else's misfortune than it is to understand a birth 'defect'. Apart from the aesthetic differences, there may also be distinctive patterns of primary socialization that distinguish the congenitally from the traumatically impaired. Many children born with disabilities have traditionally been treated as 'special' by loved ones and institutions and, as a result, may subsequently lack the social skills to actively and productively engage with the mainstream. What is clear is that within the community of Paralympic athletes, there is a hierarchy of acceptability that reflects the mainstream. This in itself is not surprising as there is bound to be a degree of political infighting surrounding the Paralympic community. Athletes who use wheelchairs as mobility aids were the first to compete in the forerunner to the Paralympic Games and a wheelchair symbol is internationally associated with 'disability' in part because it represents the type of disability to which the able bodied can relate. Therefore, as other IOSDs joined the Paralympic Movement, they were immediately accorded a lower status.

The IOSD responsible for organizing classification and competition for those athletes with mental disabilities—INAS-FID—is the newest member of the Paralympic family. INAS-FID first had athletes involved in the 1996 Atlanta Paralympic Games. Many athletes and officials from other IOSDs were unhappy about athletes with mental disabilities being eligible for the Paralympic Games. All athletes involved in the Paralympic Games to the general public have traditionally been seen as Special Olympic participants. However, as the Paralympics gained increased media exposure following the 1992 Barcelona Games, athletes began to feel they had left the stigma of being mentally impaired behind. Now there was a federation that was part of the Paralympic family that had as an entry requirement having a mental disability. Even though the hierarchy of disabilities exists (Sherrill and Williams, 1996) for athletes with physical disabilities, the stigma associated with mental disability is such that the IOSDs have found it difficult to openly embrace the participation of INAS-FID athletes (Howe, 2008).

The athlete-led political activism targeted at excluding INAS-FID athletes and committee members did not last long. A major focus of criticism for this group was the nature of the classification system. To be classified as an athlete for INAS-FID, a person must be certified by a registered clinical psychologist as having an IQ equal to or less than 75. There were fears from the early 1990s that this classification system was impossible to police because physical performance was not a determining factor (BBC, 1994). Following the Sydney Paralympics in 2000, these fears were realized when athletes from the Spanish Intellectually Disability basketball team were found to be ineligible when a team member (Carlos Ribagorda) went to the press (IPC, 2000). According to the IPC Investigation Commission:

> The process of assessment and certification of athletes with an intellectual disability for the Sydney 2000 Paralympic Games had not been properly carried out ... The IPC holds INAS-FID responsible and accountable for these violations.
>
> (IPC, 2001: 1)

While INAS-FID is still considered part of the Paralympic family, robust classification tests have yet to be developed to enable eligible athletes to compete in the next games in London in 2012. Having been excluded from the Games in both 2004 and 2008, there is mounting political pressure for INAS-FID athletes to be included in 2012 Paralympic Games (politics.co.uk, 2008). Whether these athletes make the starting line in London 2012 or not, some within the Paralympic Movement will feel hard done by, while others will feel vindicated.

The dilemma surrounding INAS-FID eligibility systems highlights the political tensions surrounding classification. On the one hand, many physically impaired athletes consider athletes with mental disabilities as not 'real' Paralympians because of their lack of physical disability. However, those who want to make the Paralympics more high profile see the aesthetically pleasing bodies possessed by INAS-FID athletes as a bonus to the movement. Yet the problems of classification of mental disability highlighted above illuminates the importance of a robust and equitable classification system.

Due to the importance placed upon equitable competition, the disability-specific classification systems for athletes with physical disabilities may create competitive pools that produce insufficient numbers of competitors to meet the IPC regulations for viable competition. The process of classification itself is grounded in medical interpretation of the functioning of the human body and, though scientific in nature, is open to abuse both by athletes and classifiers alike. Competitors are initially classified by a medical physician, a sports technical official (someone with knowledge of the physical requirements of the particular sport) and often a physiotherapist. Once classified, the athlete is allowed to compete in their designated class and will be monitored through a series of international competitions in order to confirm the correct classification. As the value of international success escalates, athletes may try to cheat the system or be declassified out of a given class because they fail to meet minimal disability criteria (Sportsmail Reporter, 2008).

Within the current climate surrounding the Paralympic Games, the collapsing of classifications into one viable event due to low numbers can be politically problematic. Events with a small number of competitors have been placed under considerable strain as a result. The problem of low numbers of competitors is exacerbated by the onset of injury in an already small pool of athletes (Howe, 2006). Eliminating a class based upon the viability of an event could be argued to be not only detrimental to future athletes who may be eligible for that category but also to lead to their disempowerment. The cancellation of an event altogether or, in some cases, the moving of competitors to a less impaired class in order to make the event viable, has an impact on future programmes. A competitor who is moved to a less impaired class is not competing on a level playing field and is unlikely to win.

Although winning is not central to Paralympic values, it is a major consideration for National Paralympic Committees (NPCs) when making team selections. NPCs emphasize winning because they receive greater publicity and increased funding based upon their position in the medal table. Individual nations are not, therefore, concerned whether events are removed from the programme unless they have athletes who are potential medallists. Events disappear from the Paralympic programme and from future programmes because of the apparent disinterest of those in the relevant classification grouping when in fact it is not solely athletes making the decisions but their NPC in conjunction with the IPC.

CONCLUSION

Progress towards a fully commercialized sporting spectacle that culminates in the quadrennial Paralympic Games has been slow in part because of the complex classification system employed, itself the product of political struggles between the various IOSDs and the IPC. The current classification systems used in sport for the disabled cause political tensions between the IPC and its constituent members because, from the IPC's point of view, they confuse spectators and deter sponsors and media interest. To this end, the IPC has in recent years tried to wrestle away control of the classification system from the IOSDs. This move and the marketing of the Olympics and Paralympics as a single entity may have undermined the IPC's autonomy to use the Paralympic Games to educate the public about athletes with a disability. The erosion of this educational imperative is problematic because one of the IPC's explicit aims is the effective and efficient promotion of elite sport for the disabled for the bettering of opportunities for the Paralympic Movement.

Decisions regarding the viability of sports competition may not necessarily be congruent with decisions about the interests of athletes with a disability in particular and disabled people in general. These decisions are by their very nature political as they are related to the balance of power in relationships between the various constituent elements within the Paralympic Movement and those who have influence over them, including the IOC. The provision of competitive categories that maximize participation may satisfy the inclusive aims of the IOSDs, but may undermine the IPC's desire to provide contests for highly motivated and skilled elite athletes. In addition, the talent pool may be spread too thinly across too many events. Disability advocacy groups are continually arguing for the rights of the disabled but, in creating an understanding of the politics of disablement, this community is articulated as homogeneous and disembodied. Paralympic sport highlights the politics of disability as a struggle between the various IOSDs and the IPC. This makes it challenging to establish a core ideology on which a majority can agree. The IPC in time may actively articulate what it means by Paralympism and certainly a robust yet still under-developed classification protocol will be at its core.

REFERENCES

Anderson, J. (2003) 'Turned into Taxpayers: Paraplegia, Rehabilitation and Sport at Stoke Mandeville, 1944–56'. *Journal of Contemporary History* 38(3): 461–75.

BBC (1994) *On the Line: Disability for Dollars*. Video shown 11 July.

Campbell, J. and M. Oliver (1996) *Disability Politics: Understanding Our Past, Changing Our Future*. London: Routledge.

De Coubertin, P. (1956 [1935]) 'The Fundamentals of the Philosophy of the Modern Olympics'. *Bulletin de Comité International Olympique* 56: 52–54.

DePauw, K. (1997) 'The (In) Visibility of DisAbility: Cultural Contexts and "sporting bodies"'. *Quest* 49(4): 416–30.

Doll-Tepper, G. (1999) 'Disability Sport', in J. Riordan and A. Krüger (eds), *The International Politics of Sport in the Twentieth Century*. London: E&FN Spon, 177–90.

Goodman, S. (1986) *Spirit of Stoke Mandeville: The Story of Ludwig Guttmann*. London: Collins.

Howe, P.D. (2006) 'The Role of Injury in the Organization of Paralympic Sport', in S. Loland, B. Skirstad and I. Waddington (eds), *Pain and Injury in Sport: Social and Ethical Analysis*. London: Routledge, 211–25.

——(2008) *The Cultural Politics of the Paralympic Movement: Through the Anthropological Lens*. London: Routledge.

International Olympic Committee (2007) *Olympic Charter*. Lausanne, Switzerland: IOC.

IPC (2000) *IPC Calls for Full Investigation*. Media release, 27 November.

——(2001) *The Paralympian: Newsletter of the International Paralympic Committee*. No. 1.

——(2003) *The Paralympian: Newsletter of the International Paralympic Committee*. No. 3.

Jones, C. and P.D. Howe (2005) 'The Conceptual Boundaries of Sport for the Disabled: Classification and Athletic Performance'. *Journal of Philosophy of Sport* 32 (2): 133–46.

Labanowich, S. (1988) 'A Case for the Integration of the Disabled into the Olympic Games'. *Adapted Physical Activity Quarterly* 5(4): 264–72.

Oliver, M. (1990) *The Politics of Disablement*. London: Macmillan.

politics.co.uk (2008) *Mentally Disabled May not be Allowed into London Paralympics*. www.politics.co.uk/news/opinion-former-index/culture-media-and-sport/mentally-disabled-may-not-be-allowed-into-london-paralympics-$1241126.htm (accessed 16 October 2008).

Scruton, J. (1998) *Stoke Mandeville: Road to the Paralympics*. Aylesbury: The Peterhouse Press.

Sherrill, C. and T. Williams (1996) 'Disability and Sport: Psychosocial Perspectives on Inclusion, Integration and Participation'. *Sport Science Review* 5 (1): 42–64.

Sportsmail Reporter (2008) *Home Malone: Irish Cerebral Palsy Footballer Kicked Out of Paralympics for not Being Disabled Enough*. 11 September, www.dailymail.co.uk/sport/olympics/article-1054524/Cerebral-palsy-footballer-kicked-Paralympics-hes-disabled-enough.html (accessed 8 December 2008).

Zola, I. (1981) *Missing Pieces: A Chronicle of Living with a Disability*. Philadelphia. PA: Temple University Press.

FURTHER READING

Bailey, S. (2007) *Athlete First: A History of the Paralympic Movement*. Chichester: Wiley.

Brittain, I. (2009) *The Paralympic Games Explained*. Oxon: Routledge.

Johnson, R. (2009) *Paralympic Sports Events*. Jefferson, NC: McFarland & Co Inc.

The Olympics and terrorism

DAVID L. ANDREWS, JAIME SCHULTZ AND MICHAEL L. SILK

INTRODUCTION

When considering the relationship between terrorism and the Olympic Games, one's attention is almost unavoidably drawn to the 1972 Summer Games in Munich. This is wholly understandable given the global media's highly emotive and enduring rendering of the events of September 1972. This process was initially realized and managed through the words of commentators such as David Coleman (BBC, United Kingdom), and Peter Jennings and Jim McKay (ABC, USA), which accompanied the chilling, if compelling, images of Black September terrorists wearing balaclavas patrolling the balcony of the Olympic Village apartment in which Israeli hostages were being held. In the recent past, the place of the Munich massacre within the global popular imagination has been re-affirmed through the influence of Kevin McDonald's Oscar-winning documentary *One Day in September* and Steven Spielberg's feature film *Munich* (2005), the former providing a detailed narration of the unfolding tragedy, the latter focusing on the Israeli state's response to the violence in Munich.

By no means wishing to discount the significance of Munich as the most 'obvious manifestation' (Wedemeyer, 1999: 217) of terrorism within the Olympic arena—it will, after all, be discussed at some length within this chapter—it would be equally remiss to focus on it exclusively. This is because the relationship between terrorism and the Olympic Games is more diverse, and indeed complex, than one would immediately anticipate. Thus, our aim in this chapter is to complicate popular understandings of terrorism, through recourse to its various, and varied, Olympic manifestations. This process of complicating terrorism is centred on a recognition of both violent acts against the state, and violent acts perpetrated by the state, as examples of terrorism, broadly understood. Both these forms of terrorism which have, at various times, been materialized within the context of the modern Olympic Games since its inception in 1896, will be discussed.

Before embarking on such an explication, it is first necessary to clarify precisely what is meant by terrorism. In the broadest terms, it is possible to refer to terrorism as acts which are violent, or which carry the threat of violence, and are designed to create a pervasive and invasive climate of fear and anxiety among the general populace. Such acts are usually (although by no means always) perpetrated by individuals, or groups of individuals, belonging to organizations motivated by stated objectives, be they political, criminal, or eccentric (Lutz and Lutz, 2004). Furthermore, terrorist tactics can either be aimed at high-profile representative figures or at the general population. In terms of the former, attacks against political, economic, religious, or military leaders are frequent occurrences since these individuals act as representative subjects of the society targeted by the terrorist action. This *monumental terrorism* is motivated by the desire to disrupt the very foundations of a society, by directly attacking the literal embodiments of the core institutions (political, economic, religious, or military) around which the general populace orders its very existence. Thus, while the loss of life may be relatively small in scale, its desired significance and effects are expansive. Conversely, *vernacular terrorism* eschews the targeting of a society's leaders, in favour of the seemingly indiscriminate targeting of the anonymous and the innocent as they conduct their everyday lives. The logic underpinning such acts—which often result in far greater numbers of fatalities and injuries—is to encourage people to perceive themselves as being under constant threat of attack. Therefore, whether it be targeted at high-profile figures (*monumental*), or the general populace (*vernacular*), the aim of terrorism is undifferentiated: to create a pervasive climate of disquiet and instability which could lead to the realization of particular aims and objectives. Of course, some acts of terrorism are motivated by broader political concerns and strategies. For instance, terrorist strikes have been motivated by the desire to create a swift and severe backlash from the state, thereby accentuating the authoritarian and undemocratic tendencies against which the terrorist perpetrators were agitating, and around which the terrorists look to build more oppositional support.

What has been described thus far is the conventional understanding of terrorism in which violence—deriving from a non-state entity, and contrary to the normative principles of ethical behaviour operating in the society in which it is enacted—is utilized in an attempt to destabilize an established social formation. Complexities arise because within virtually all societies there exist varying degrees of (dis)agreement regarding the rectitude of any social order, its attendant hierarchies and constitutive practices. In any society, let alone between societies, the very practice of terrorism, and the subjectivity of the terrorist, are highly contested terms whose meaning and attribution are dependent on elements such as the cultural and economic location of those comprehending the 'terrorist' act. Differently put, 'terrorism, like beauty, is in the eye of the beholder' (Lutz and Lutz, 2004: 8). Acts of political violence tend to be considered terrorist when they are aimed at and/or within states whose system of governance is broadly accepted by the general public. The terrorist ascription is, however, less clear cut within societies fraught by civil strife and internal dissension, i.e. formations in which a consensual system of rule has yet to be established, and in which competing factions frequently engage in violent power struggles. Furthermore, although most understandings of terrorism consider it to refer to acts of politically-motivated violence against a state formation, some commentators have identified the state as a potential perpetrator of terrorist acts as well. According to such thinking, state terrorism utilizes violence, and the threat of violence, against its citizenry in order to create a climate of widespread paranoia and insecurity aimed at defusing any oppositional tendencies, and thereby stabilizing the authoritarian order.

TERRORISM AGAINST THE *OLYMPIC* STATE

Given the status of the Olympic Games as an unequivocal global 'mega-event' that has come to capture the attention of the world's viewing public (Roche, 2003), it is perhaps surprising that the Games have thus far been the target of relatively few terrorist operations. This can possibly be attributed to the Olympic Movement's carefully crafted—if contradictory and romanticized—ethos of internationalism, humanism and pacifism, which for much of the 20th century allowed the Games to exist and operate in a space somehow dislocated from the harsh realities of global tumult. Of course, the Olympics as a modernist initiative, and indeed the various Olympiads through which the movement became actualized, have never been separated from their socio-political context (as the games of Berlin in 1936, and Mexico City in 1968, graphically attest). Indeed it would be an impossibility for them to do so. Nevertheless, the appearance of Olympic autonomy appeared to persist within global popular consciousness, arguably until it was conclusively interrupted on the morning of 5 September 1972, at the Games of the 20th Olympiad held in Munich, West Germany.

The Munich Games had been painstakingly designed to distance Germany from its recent Nazi past. The aim was to stage 'The Games of Peace and Joy' that would provide an Olympic and ideological counterweight to the fascist militarism of the Berlin Olympics held 36 years earlier and conceived as a platform for exhibiting the political, cultural and racial supremacy of Hitler's Germany. As the West German Chancellor, Willy Brandt, later lamented, the Munich Games were supposed to 'go down in history as a happy occasion' (Brandt, 1978: 439). Instead, they would become synonymous with the incursion of terrorism into the sporting world, as a result of the murderous actions of the Black September terrorist organization during that 'one day in September' (Reeve, 2000)—hostage taking, a resultant siege, and a failed rescue imbroglio that ultimately led to the deaths of 11 members of the Israeli Olympic team, one West German police officer and five members of the terrorist group.

The Black September organization, an extreme faction of the Palestinian Liberation Organization (PLO), was motivated by the goal of ending the Israeli occupation of Palestine and restoring a Palestinian homeland. The group's name derived from the expulsion of the PLO from Jordan in September 1970, a move that provoked a number of terrorist attacks by the PLO against Jordanian and Israeli targets. Although the conception of Black September's Munich plot can be seen as an extension of its broader struggle with the Israeli state, another contributory factor proved to be the International

Olympic Committee's (IOC) refusal to allow Palestinian participation in the Olympics. According to the leader of the terrorists, Luttif Afir (known as 'Issa'), the 1972 Games provided the Black September faction with a spectacular opportunity for showcasing Palestinian 'grievance(s) to the millions watching around the world' (Reeve, 2000: 51). Furthermore, while involving a sizeable number of fatalities, this act could be considered an example of *monumental terrorism*. The Black September group explicitly targeted Israeli athletes and officials, while releasing unharmed those from the Hong Kong and Uruguayan teams housed in neighbouring apartments. These members of the Israeli Olympic team were prized as surrogates of the Israeli state, and subsequently used as embodied capital in ultimately fruitless negotiations, through which the terrorists sought to realize their numerous demands (the most significant of which being the release of 234 Palestinians from prisons in Israel, in what would have been tantamount to the trading of the embodiments of one geo-political formation for those of another).

The Israeli state's unwavering refusal to negotiate with the terrorists—in Prime Minister Golda Meir's words, 'If we should give in … then no Israeli anywhere in the world can feel that her life is safe' (quoted in Reeve, 2000: 61)—coupled with the terrorists' increasingly volatile state, and a series of miscalculations and errors of judgement by the German authorities, ultimately led to the bloody denouement of this Olympic crisis on the tarmac at Fürstenfeldbruck military airport. Thus, with one brutal incident, the view of the Olympic Games as a sanctum of internationalism, humanism and pacifism somehow distinct from an increasingly fractious, inhumane and violent world, was conclusively shattered. As IOC President, Avery Brundage, noted in his notorious speech at the memorial service held a few hours after the killing had ceased, 'Sadly, in this imperfect world, the greater and the more important the Olympic Games become, the more they are open to commercial, political, and now criminal pressure' (quoted in Reeve, 2000: 149). However, the importance of the massacre extended beyond the confines of the Olympic Games. The uncomfortably intrusive, yet clearly compelling, television coverage of the hostage crisis was communicated to a watching world that witnessed this 'public killing event' in real time (Tulloch, 2000: 225). As a consequence, and as terrorism expert Bruce Hoffman noted, the events at Munich were 'the most consequential terrorist incident in history prior to 9/11' (quoted in Gilgoff, 2004: 39).

The global exposure afforded the Black September terrorist operation changed the manner in which the Olympic Movement perceived itself, and was viewed by the external world. No longer simply responsible for a celebration of sporting and human achievement, the IOC and host organizers now concern themselves with countering what has become an ever-present terrorist threat, while a number of extremists (groups and individuals) have looked to exploit the global spectacle that is the contemporary Olympic Games for their own gain (as illustrated by threatened, but not realized, terrorist actions at both the 1992 Barcelona and 2000 Sydney Games).

To date, the last terrorist act, as conventionally understood, to take place at an Olympic venue was the explosion of a pipe bomb at the Atlanta Games' Olympic Centennial Park on 27 July 1996, which led to two deaths and 111 injuries. The long-time suspect for the bombing was a security guard, Richard Jewell, whose actions in alerting the police and evacuating the area caused him to be initially hailed as a hero. However, the perpetrator turned out to be the notorious fugitive, Eric Rudolph, who had carried out a series of bombings across the American south, motivated by his peculiar synthesis of anti-abortion, anti-homosexual and anti-corporate sentiments. In a 2005 statement released following his guilty plea to four bombings, including that at the Atlanta Games, Rudolph provided insights into his choice of the Olympics as a bombing venue:

> For many years I thought long and hard on these issues and then in 1996 I decided to act. In the summer of 1996, the world converged upon Atlanta for the Olympic Games. Under the protection and auspices of the regime in Washington millions of people came to celebrate the ideals of global socialism. Multinational corporations spent billions of dollars, and Washington organized an army of security to protect these best of all games. Even thought [sic] the conception and purpose of the so-called Olympic Movement is to promote the values of global socialism, as perfectly expressed in the song 'Imagine' by John Lennon, which was the theme of the 1996 Games—even though the purpose of the Olympics is to promote these despicable ideals, the purpose of the attack on July 27th was to confound, anger and embarrass the Washington government in the eyes of the world for its abominable sanctioning of abortion on demand.
>
> (Rudolph, 2005)

As eccentric as Rudolph's personal politics may have been, his terrorist strategizing displayed a cogent understanding of the cultural—and, by extension, the political—value of the Olympic Games. The target of this domestic terrorist's action was not innocent spectators, or representative figures such as athletes or Games officials. Rather, Rudolph focused on causing major disruption to the Games themselves. Originally, he had hoped to force the cancellation of the Games, 'or at least create a state of insecurity to empty the streets around the venue and thereby eat into the vast amounts of money invested' (Rudolph, 2005). In the first instance, Rudolph had intended to achieve his goal by bombing Atlanta's power grid. Once this proved unfeasible, he moved on to planning a series of five smaller-scale explosions on consecutive days at the beginning of the Games, that targeted various Olympic locations. In his 2005 statement, Rudolph claimed that, in the hour leading up to each explosion, he had intended to warn the authorities of the location and the time of detonation of the explosive device, thereby providing an opportunity for clearing the area of 'innocent civilians', and 'leaving only uniformed arms-carrying government personnel exposed to potential injury'. The indiscriminate killing and injury caused by the Centennial Park 'disaster' (Rudolph, 2005) proved antithetical to Rudolph's (admittedly post facto) stated rationalization of his motives, and caused him to target more narrowly in subsequent attacks against health clinics and homosexual nightclubs.

TERRORISM BY THE *OLYMPIC* STATE

If terrorism can be understood as the act or threat of violence against an identifiable population—motivated by broader aims than the killing or maiming of victims, be they random or representative targets—then it soon becomes evident that both state and non-state actors have been culpable in this regard. In other words, terrorism is not simply the domain of non-state insurgents looking to disrupt the equilibrium of the state formation. Indeed, there are numerous examples of the state involving itself in operations designed to either disrupt (most effected in external settings) or stabilize (the usual practice in internal settings) the social order through the utilization or threat of violence. Of course, state terrorism, unlike its non-state counterpart, is able to operate with a degree of impunity, particularly with regard to its internal operations, because the 'strong' state is able to claim a monopoly on 'legitimate' violence, and is able to set the boundaries which enforce this monopoly (Smelser, 2007: 242). Thus, the state can operate as a coercive actor, and certainly the history of the modern Olympic Games provides numerous examples of the hosting of an Olympiad becoming the occasion for the exercise—or perhaps more accurately the amplification—of state terrorist initiatives against the host populace, or sections thereof.

The reign of terror enacted by the militarized Nazi Party in Germany during Hitler's Third Reich certainly framed the manner in which the 1936 Berlin Olympics was structured and presented to the German people. The sporting prohibitions and restrictions meted out against German Jewish athletes (Guttmann, 2002) in the preparation for, and actualization of, the Berlin Games may seem inconsequential when positioned against the ravages of Kristallnacht, or the horrors of Auschwitz-Birkenau. However, they are surely related and contributory elements in the Nazi State's relentless march towards Jewish genocide.

The events surrounding the violent suppression of student protest in the weeks leading up to the 1968 Mexico Olympics offer perhaps a more obvious, and certainly more generally acknowledged, example of state-generated terrorism directed against constituents of a host population. In terms of fatalities, it was certainly the most murderous. As Hoberman noted, Mexico had successfully sought the right to host the Olympic Games in 'order to gain international recognition of its transformation into a modern or semi-modern country'. However, in the 'context of the student revolt and the repression that ensued, these celebrations seemed nothing but gaudy gestures designed to hide the realities of a country stirred and terrified by governmental violence'

(Hoberman, 1986: 13–14). As 1968 unfolded, as in many other countries, a wave of student-led popular protest had spread throughout Mexico, mobilized around demands for social justice, democratic reform, an end to governmental corruption and the release of political prisoners. As the protests became more forceful, so President Gustavo Díaz Ordaz's response became more retaliatory and aggressive, culminating in the Plaza de las Tres Culturas (also known as Tlatelolco) massacre of 2 October 1968, 10 days before the opening of the Olympic Games.

The spectacle of the Games proved to be a focal point for the protesters eager to concentrate the nation's, and indeed the wider world's attention, on its various grievances. Concomitantly, the long-time incumbent *Partido Revolucionario Institucional* (PRI) Government sought to use the Games as a showcase for its 'state-promoted "Mexican miracle" of rapid industrialization and urbanization', and therefore looked to suppress any form of visible social dissent and disharmony (Frazier and Cohen, 2003: 623). The consequence of the inevitable impasse between a repressive government and protesting students was all too predictable. As 2 October drew to a close, more than 10,000 protesters congregated in the Plaza de las Tres Culturas, listening to rallying speeches and chanting provocative slogans such as 'No Queremos Olimpiadas, Queremos Revolucion!' ('We don't want the Olympics, we want revolution!'). They were about to disperse when the attendant force of more than 5,000 heavily armed soldiers (amongst whom were the Olympia battalion, an elite paramilitary group charged with Olympic security) opened fire on the defenceless throng. The result was a death toll estimated at anywhere between 35 and more than 1,000, with hundreds more wounded and/or arrested. Within a few bloody hours, the Díaz Ordaz regime crushed the student protest movement and re-imposed public order in time for the all-important execution of the Games. Thus, state terrorism was responsible for what has been described as the 'worst crime in Olympic history'(Hoberman, 1986: 1).

In the lead up to the 1988 Seoul Olympic Games, there was every indication that a repeat of the state-perpetrated violence of some 20 years previous was a distinct possibility. The Games occurred during a time of great social upheaval within the Republic of Korea (South Korea), the intensifying unpopularity of President Chun's military-backed regime prompting the growth of political opposition movements clamouring for constitutional reform. Within this context, the Seoul Games were viewed by many South Koreans with distrust—and even disdain—because they were regarded as a high-profile expression of the Chun regime's 'Sport Republic' policy. This political strategy sought to mobilize sporting events and successes as a means of legitimating the Chun administration, and the position of leadership 'in the eyes of both the people of Korea and the international community' (Ha, 1998: 11). However, the Games brought with them not inconsiderable political problems related to the exacerbation of tensions with the Democratic People's Republic of Korea (North Korea), prompted by the staging of this global event in Seoul, the possibility of a Soviet bloc boycott of the Games as a show of solidarity with North Korea and, most pertinent to this discussion, the transition to a constitutional democracy which was promised before the beginning of the Games. The situation became ever more fraught in April 1987, following President Chun's suspension of the debate on constitutional reform until after the Seoul Games. This precipitated a re-awakening of the student-led protest movement which mobilized on to the streets of South Korea's major cities. While much of this political protest dissipated following Roh Tae-Woo's (the person chosen as Chun's immediate successor) 29 June Declaration, amongst the more radical factions of the student movement, there continued to be smouldering resentment toward the Games, suggesting the very real threat that they would be the site for pro-democracy demonstrations. In response to this possibility, the Seoul Government flexed its authoritarian muscles both by highlighting the scale and aptitude of the security forces on operation during the Games, and by turning many parts of Seoul into an 'Olympic Peace Zone' in which anti-government protest was forbidden under threat of immediate arrest (Chira, 1988). Suppressed by this symbolic and material government intervention, and neutered by a tide of popular support for the Games that swept the nation, the Seoul Olympics passed without any further discernible expression of social unrest.

Parallels can be drawn between the repressive climates that provided the backdrop for the 1968 and 1988 Olympics, and that which enveloped the 2008 Beijing Games. This is not to suggest that the levels of violence witnessed in Mexico City were in any way replicated in Beijing. Rather, and more analogous to the events surrounding the Seoul Games, the Beijing Olympics provided the platform for the operation of what continues to be an authoritarian Chinese State. Akin to many previous Olympiads, the Beijing Games were viewed as an opportunity to explicate, and indeed celebrate, the nation's progress: in this case, the Olympics being a showcase for the social, economic and technological advances made by the People's Republic of China in becoming a truly global power. In order to realize the goal of presenting the best possible Games and thus the best image of China, the Chinese Government poured billions into capital investment (creating spectacular and truly iconic built environments) and elite performance programmes (with the aim of ensuring significant athletic success). The former did not come without a human cost, with an estimated 1.5m. forced evictions and relocations prompted by Olympic venue development and general beautification schemes. The Chinese Government also engaged in more clandestine practices, designed to suppress dissent in the name of the enforced social conformity and stability required to present an exemplary Games—and, by implication, an exemplary society—to the watching world. Thus, press freedoms were curtailed, the internet even more heavily censored, and the systematic detention, intimidation or harassment of anti-government activists elevated to new levels. Furthermore, in May 2006, the Government introduced the 'Re-education Through Labour' scheme, whereby individuals could be detained in compulsory labour camps without being legally sentenced for up to four years for minor offenses including vagrancy, begging and peaceful protest (www.amnesty.org). This punitive measure, clearly designed to remove from Beijing's constituent populace undesirables (the overtly poor and politically oppositional), significantly compromised Chinese human rights. Of course, the Beijing Olympics were not unique in this regard, with similar practices of forced relocation (and, not infrequently, imprisonment) for contextually contingent 'undesirable' subjects having reportedly occurred during the build up to Berlin (1936), Moscow (1980, of the disabled and poor), Los Angeles (1984, of gang members), and Athens (2004, of the homeless, addicts and asylum seekers).

It is also important to note how the Olympic state is able to strategically mobilize the public panic embracing the spectre of terrorism to advance its own political ends. This was plainly evident with the Beijing Olympics. The Chinese Government engaged in overt forms of social repression to ensure the Games were delivered in a suitably controlled manner. Herein, it could be argued that China utilized the (*monumental*) spectacle of the Olympic Games as a catalyst for creating a climate of paranoia and insecurity aimed at defusing any oppositional tendencies, and thereby stabilizing what is in reality a more explicitly authoritarian social order than those evident within Western liberal democracies. As Sugden (2008) noted, there is a widespread belief that 'China has used the Olympics and the accompanying "threat of terrorism" as a pretext for cracking down on civil rights protesters and ethnic minority groups'. For this reason, some may argue that the Beijing Olympics were indeed both an expression and an agent of state terrorism. However, as Chomsky identified, the attribution of the label 'terrorist' is steeped in the moral relativism of those doing the attribution. Hence, for some, the Beijing Olympic enterprise represents little more than a natural, and legitimate, expression of China's system of government.

The subjective nature of the terrorist label was evidenced within the Beijing Olympic context, specifically in regard to the political mobilization of the Uighur question. To some, the Uighurs are an oppressed Muslim ethnic minority predominantly located in East Turkestan, in the western region of China, whose legitimate struggle for independence has been brutally crushed by the Chinese Government. To others, notably the Chinese State, the Uighur movement represents a dangerous and destabilizing threat to Chinese stability. Although the clamour for Uighur independence has been largely expressed in non-violent terms, the various terrorist acts perpetrated by the more militant strands of the Uighur separatist movement were used as a pretext by the Chinese Government for the suppression—some would say, repression—of individual rights of protest at, and around, the Olympic site.

That said, whilst it is easy to castigate the workings of non-Western governments, arguably an equally insidious, and ultimately catastrophic example of the state's mobilization of terrorism-induced anxiety among the general populace was provided by the 2002 Salt Lake City Winter Olympic Games. Coming as it did a few months after the 11 September 2001 terrorist attacks in the USA—and at a moment of heightened American sensitivity to its status as a nation 'under threat'—the carefully choreographed and highly emotive nationalism that dominated the Games' opening ceremony cannot be divorced from the Bush Administration's desire to secure a groundswell of popular support for what were, at that time, already pre-ordained retaliatory incursions into Afghanistan and Iraq (Silk and Falcous, 2005). In this case, the Olympic Games were utilized as a means of advancing the spectacle of terrorism by a state looking to exact its own regime of retributive terror. Such, then, are the complexities of the relationship between the Olympics and terrorism, and indeed the indivisibility of geopolitical and sporting conflict in the contemporary age.

CONCLUSION

While conventional conflict between warring states remains a part of the geopolitical landscape, there is little denying that the last 40 years have witnessed the spread of politically motivated terrorist actions against the representative institutions and embodiments of nation-states. This trend was confirmed, and indeed catalysed, in the eyes of the global populace by terrorist events in the USA on 11 September 2001, in Madrid on 11 March 2004, and in London on 7 July 2005. Although the London bombings were arguably most closely related to the Olympic Games (coming as they did, the day after the city was announced as the successful bidder for the 2012 Games), the global climate of terrorist-anxiety created by these actions, and frequently mobilized by national governments looking to introduce new law and order directives, has seen the general relationship between terrorism and the Olympics become more readily apparent. Indeed, terrorism has become an unavoidable element of the Olympic discussion. No longer merely subject to the perils of financial risk, now the complexity of the Olympic Leviathan renders it equally exposed to the risk of terrorist action (Jennings, 2005). As a consequence, terrorism is now an important aspect in the risk management strategizing of any Olympic Games, or, indeed, any Olympic bid. For instance, the IOC purchased US $170m. of insurance for the 2004 Athens Olympics, as protection against acts of terrorism or natural disasters that would disrupt or cancel the Games. Underscoring the new normality of the terror-Olympic relation, as host cities are named, an important point of debate would appear to be who precisely offers the main potential terrorist threat.

Furthermore, the 'legacy' for the nation that the IOC now mandates for successful bidding cities could yet assume some previously unanticipated forms within the present climate of fear with regard to the spectre of terrorism. As well as sporting, housing, transport and communications infrastructure, future Olympic Games may well bequeath to their host cities the sophisticated and invasive surveillance infrastructure that has come to be viewed as an indispensable requirement of a stadium that is suitable for an Olympic Games. Indeed, the costs for Olympic security now represent one of the most significant single items in the Olympic budget, with the curiously titled seven-year long 'Grand Beijing Safeguard Sphere' programme costing an estimated $6,500m. (Sugden, 2008). However, such initiatives have worrying implications for the civil liberties of those living within the newly technologically fortified Olympic cities. As Sugden has warned:

> ... an event like the 2012 Olympics may be used by government and security-related interest groups to magnify in the public mind the terrorist threat and construct a 'climate of fear' to justify technologically driven control strategies, to counter anti-social behaviour and democratic protest, to exclude the dangerous 'others' from public space, and to introduce identity cards that link citizens to state held databases.

If that is indeed the case, there is every possibility that the terrorism-induced anxiety that has come to envelop the Olympic Games will result in authoritarian states being the only ones able, or indeed willing, to stage a 'safe' Olympics. Alternatively, the lure of hosting the Games could encourage democratic states towards more authoritarian systems and practices of governance. Whichever it may be, the relationship between terrorism and the Olympic Games would now appear to be permanent.

REFERENCES

Brandt, W. (1978) *People and politics*. Boston, MA: Little, Brown and Co.

Chira, S. (1988) 'Alone in dissent in Korea: Although student protests set the agenda, this year they fail to gain wide backing'. *New York Times,* 17 August.

Frazier, J.L. and D. Cohen (2003) 'Mexico '68: Defining the Space of the Movement, Heroic Masculinity in the Prison, and "Women" in the Streets'. *Hispanic American Historical Review* 83(4), 617–60.

Gilgoff, D. (2004) 'The meaning of Munich'. *US News and World Report*, 14 June, 39.

Guttmann, A. (2002) *The Olympics: A History of the Modern Games*. Urbana, IL: University of Illinois Press.

Ha, W.-Y. (1998) 'Korean sports in the 1980s and the Seoul Olympic Games of 1988'. *Journal of Olympic History*, Summer: 11–13.

Hoberman, J. (1986) *The Olympic Crisis: Sports, Politics, and the Moral Order*. New York: Caratzas Publishing.

Jennings, W. (2005) 'London 2012 and the Risk Management of Everything Olympic'. *Risk & Regulation,* Winter(10), 7–9.

Lutz, J.M. and B.J. Lutz (2004) *Global Terrorism*. London: Routledge.

Reeve, S. (2000) *One day in September*. London: Faber & Faber.

Roche, M. (2003) 'Mega-events and media culture: Sport and the Olympics', in D. Rowe (ed.), *Critical Readings: Sport, Culture and the Media*. Berkshire: Open University Press, 165–81.

Rudolph, E. (2005) Full text of Eric Rudolph's confession. 14 April, NPR, www.npr.org/templates/story/story.php?storyId=4600480.

Silk, M. and M. Falcous (2005) 'One day in September/A week in February: Mobilizing American (Sporting) nationalisms'. *Sociology of Sport Journal*, 22(4): 447–71.

Smelser, N.J. (2007) *The Faces of Terrorism: Social and Psychological Dimensions.* Princeton, NJ: Princeton University Press.

Sugden, J. (2008) 'Watching the games'. *Foreign Policy in Focus*, 22 August, www.fpif. org/fpiftxt/5490.

Tulloch, J. (2000) 'Terrorism, "killing events," and their audience: Fear of crime at the 2000 Olympics', in K. Schaffer and S. Smith (eds), *The Olympics at the Millennium: Power, Politics, and the Games.* New Brunswick, NJ: Rutgers University, 224–42.

Wedemeyer, B. (1999). 'Sport and terrorism', in J. Riordan and A. Kruger (eds), *The international politics of sport in the twentieth century.* London: E&FN Spon, 217–33.

FURTHER READING

Marcovitz, H. (2002) *The Munich Olympics.* New York: Chelsea House Publishers.

Richards, A., P. Fussey and M. Silk (eds) (2009) *Terrorism and the Olympics.* London: Routledge.

Stephens, A.C. and N. Vaughan-Williams (eds) (2008) *Terrorism and the Politics of Response: London in a Time of Terror.* London: Routledge.

Berlin 1936

CHRISTOPHER YOUNG

INTRODUCTION

The 1936 Olympics are by some margin the most written about in the history of the modern event, yet their historiography has been tarnished by the lure of hindsight, a continued fascination with fascism, and factual inaccuracy born of the desire to cling to convenient mythology. Both for dedicated sports specialists and mainstream historians citing them briefly in their general accounts of the period, Berlin 1936 largely remains the 'Nazi Olympics' or 'Hitler's Games'. Two narratives dominate: the hypocrisy of hosting a peaceful event as the precursor to the most devastating war in human history, and in this setting, the morality tale of Jesse Owens defying Adolf Hitler's Aryan policies with four gold medals and superlative performances in track and field. Yet such emphases are reductive. To read the Berlin Games through the lens of later events—in addition to genocide and war, plans to host a Reich version of the event in a 400,000 capacity stadium in Nuremburg, evocation of the Olympic spirit at the Eastern front, and the trenches and flak machines operated by youths on the Reichssportfeld in the final defence of the city are oft-cited examples—is to overlook the fact that they were awarded by the International Olympic Committee (IOC) (by postal ballot, in preference to Barcelona) as a gesture of reconciliation to a democratic Germany in 1931 and, after the Nazi seizure of power in 1933, executed by the regime in tandem with two leading functionaries from the world of Weimar sport (Carl Diem and Theodor Lewald, himself half Jewish) (Eisenberg, 1999; Keys, 2006). In addition, focusing excessively on Owens perpetuates the myth that Hitler refused to congratulate him (a snub, it is now well established, which never occurred), whilst the US athlete was in fact more slighted by President Roosevelt's refusal to send him a telegram and compared the adulation he received from the German public favourably with his mixed reception at home (Krüger, 2003b). Certainly, Owens is lovingly portrayed in Leni Riefenstahl's 'Olympia', a film which appeared in 1938 (to mark Hitler's birthday) and has since cast a particular fascist sheen on the general appreciation of the Games (Graham, 1986).

The double-bind of Riefenstahl's film—a work of art (crowned by international juries before the war and rehabilitated by art-house audiences in the early 1970s) which captures the beauty of a sporting event with clear propagandist intent—offers perhaps the most balanced way of considering the

Games themselves. When dealing with the 1936 Olympics, we are faced not so much with *either/or* choices but with *both/and* possibilities. It is not necessary, for instance, to view 1936 as the natural climax of a sporting ideology with proto-fascist roots (Hoberman, 1995). Whilst many IOC members had an ingrained antipathy towards Jews and came from countries whose governments had tilted towards the right, it must be remembered that the bourgeois elites that made up the committee (including those from the USA) had more to fear from Bolshevism than Nazism, and thus protested against and acquiesced with the hosts in equal measure. Hitler was told firmly on the opening day, for example, to greet all winners (not just Germans) or none at all—an instruction he followed by opting for the latter—but the IOC remained reluctant to probe the contentious racial selection policy of the German national team. Nor is it sensible to focus on the displacement of criminals, prostitutes and hundreds of gypsies from the city centre (the latter to a camp in the suburb of Marzahn, whence they were transferred to Auschwitz in 1943), the removal of anti-Semitic signs and discourse from public areas and newspapers, and the swift return of brutality at the end of the Games. Such acts were deplorable, but at the same time the organization of the Games largely in accordance with international regulations produced an Olympic pause and brief return to the atmosphere of the Weimar Republic, albeit one that had no effect on the regime's politics or the subsequent course of history. Equally, the organizers were both bound by the regime and able to preserve an important modicum of autonomy. International pressure prevented the dismissal of Diem and Lewald from the already appointed organizing committee, but its saturation with party officials and the annexation of its overseeing body (the German Olympic Committee) secured capillary control. None the less, Diem and Lewald resisted and collaborated to their best advantage, benefiting from huge financial windfalls, asking Joseph Goebbels to enhance publicity, and gaining ground in general for sport after decades of conflict with the gymnasts (Turnen) (Eisenberg, 1999). In almost every aspect, therefore, it is vital to consider the Games in the round.

Before examining the event in more detail, it is important to introduce some further caveats. The Berlin Olympics did not evolve in a vacuum, and viewing them discretely or as a simple symbol of Nazi evil can lead to mis- or over-interpretation. Contextualization is essential, and several spheres demand attention.

CONTEXT 1: OLYMPISM

The Nazis did not manipulate Olympic ideals, but gently appropriated what they found on offer. As formulated by Pierre de Coubertin, Olympism was light on definition but heavy in symbolism. The founder aimed not simply to propagate abstract ideas but to generate feelings, fantasies and experiences that would enhance individual and collective world views through the act of participation. Verbal codes were, therefore, broad-brush, and the regime—like dictatorships that followed them (Moscow 1980, Beijing 2008)—had little difficulty publicly espousing values such as world harmony and chivalrous competition. Whilst the emphasis on the experiential played to the Nazis' strengths, their desire to stimulate and galvanize public emotion merely accentuated recent Olympic practice. In the two immediately preceding Games (Amsterdam 1928, Los Angeles 1932), commentators noted the awe and respect the ritual aspects of the event had instilled.

The Olympics' ceremonial inventory had developed incrementally since 1896. The first Games, inspired by the Paris Expo of 1889, already contained an opening ceremony with flags, hymns, cannon shots, the release of doves, and raised flags to celebrate the victors; at Antwerp 1920, an oath was introduced for the athletes to swear under the new Olympic five-ring flag, which Coubertin had designed in 1914; the motto 'Citius, Altius, Fortius' was suggested by the French cleric, Henri Didon, at the 1921 IOC session; and Amsterdam introduced the Olympic flame in 1928. Los Angeles, like the Workers' Olympics in Frankfurt in 1925 (Riordan, 1999), inserted a show element into the opening ceremony and tidied up the medal presentations, with a rostrum incorporated into the event itself. Berlin added the Olympic Bell, which was not continued, and the Torch Relay, which by contrast was to become one of the Games' most potent invented traditions.

Relays featured on ancient Greek vases and Diem had organized such events in Germany, where they had been popular from the turn of the century. In 1934, 120,000 runners brought messages to Hitler in Koblenz, but the Olympic version was more modest, requiring just 3,075 runners from five different countries to bring the flame from the site of the ancient Games in Greece to the German capital. The itinerary became more lavish and ingenious thereafter. Connecting with the stadium cauldron introduced by the Dutch, publicizing the Games, symbolizing peace and understanding, and creating a visual link between ancient and modern, the relay was warmly supported by de Coubertin. It also resonated with Germany's Philhellenism (a love of Greece that was writ large over the plans to host the cancelled Games of 1916 in Berlin), although its combination with Nordic nature in Riefenstahl's film lends the event a heavier ideological tone than its inventor probably intended.

CONTEXT 2: SPORT'S POPULARITY IN THE INTER-WAR YEARS

The Nazis did not need to arouse the public's interest in sport—since this was a given by the 1930s. As in other countries, from the late 19th century, sport had competed, in Germany often bitterly, with gymnastics (Turnen), which had already established itself as an expression of nationalist body culture. Across Europe, the First World War marked a turning point in sport's popularity, and it developed a mass appeal across a wide social range in the inter-war period, as well as expanding its international structures and federations. The Nazis had taken no official position on the gymnastics versus sport debates before coming to power, although the *Völkischer Beobachter* in 1929 pushed the idea of a native German physical culture over internationalism and generally opposed participation in, and the hosting of, the Olympics. From 1933 onwards, however, National Socialism became a syncretism of all things popular, and since sport was so obviously in keeping with the modern, urban, popular mood, it was taken under the protective arm of the regime and centralized on the Italian model. Mass Turnen displays continued (not least at the 1936 Games), but Turnen was now institutionally incorporated into sport.

The Games of 1936 were very much of the moment. Even depression-hit, west-coast America had more than doubled Olympic attendance in 1932, taking it over the 1m. mark, and Berlin, at the heart of Europe, was to attract well over 3m. in 1936 (Keys, 2006). The press had covered the Games before, but in Berlin the public's interest in the event meant journalists came in droves, with radio providing 3,000 transmissions (many, for the first time, live) from state-of-the-art facilities to 40 countries. In 1932 the world's leading sports nation had raised the bar in organizational standards: the first (temporary) village had been constructed for the athletes, traffic lanes had been re-routed, and events ran like clockwork. In Berlin, the Games were hosted superlatively. A permanent Olympic Village was crafted with as much care as the main sites, ticket prices were lowered to ensure good attendance, radio and newspaper coverage was blanket, 100,000 Berliners had access to television footage, and technical innovations such as the automatic photo-finish caused great excitement. Importantly, the athletes rose to the occasion. In track and field alone, 17 world and 27 Olympic records were broken. As retrospectives (even from Jewish spectators) show, the Games were passionately enjoyed as a sports event by spectators and athletes alike.

CONTEXT 3: SPECTACLE

Berlin 1936 was the pinnacle of Olympic spectacle—one matched only by the Hollywood show of Los Angeles 1984 (a comparison that is never made) and Beijing 2008 (one that is all too readily)—but spectacle was not invented by the Nazis. In fact, the Olympics had struggled to step out of the shadow cast by the extravagant World Fairs, from which de Coubertin had drawn inspiration for the Games. At the Wembley imperial Expo of 1924–25, for instance, a self-proclaimed 'stock-taking of the whole resources of the Empire', the stadium that staged the opening and closing ceremonies already had a capacity of 100,000, the same as that in Berlin 12 years later; the rituals were attended by the monarch and the radio broadcast of the opening conducted by King George V was the first time the British public as a whole had been gathered together to participate in a national event through the new medium. Each day, the stadium was the venue for the daily 'Pageant of Empire', which involved 15,000 performers and hundreds of animals, and considerably eclipsed Diem's Festival Play, which was performed only four times in Berlin. The new mass medium of film was made accessible to a broad public, who were mainly treated to travel films from African parts of the empire. Certainly, in the realm of sport, the Socialist Workers' Sport International's Workers' Olympics in Frankfurt (1925) and Vienna (1931), as well as the Czech Workers' Gymnastic Federation events in Prague (1921, 1927, 1934) set standards that Olympic organizers could not afford to ignore: Frankfurt featured mass choirs, flag waving and mass physical artistic displays, whilst Vienna hosted 100,000 spectators in a purpose-built stadium (Roche, 2000). If Los Angeles had set one mark in 1932, then bourgeois sport's ideological rivals gave the 1936 organizers an equally ambitious one at which to aim as well.

CONTEXT 4: SPORT AND POLITICS

Whilst it is a truism to call the 1936 Games a political event, the Olympics, and indeed sport in general, had been politicized long before their encounter with National Socialism. The inaugural Games of 1896 caused the collapse of two Greek governments, the British and Americans clashed in 1908 over the Irish question, the Finns were allowed to compete in their own right in 1912 despite Russian occupation, and the Germans themselves had been excluded from the two Games immediately after the First World War (Antwerp 1920, Paris 1924). National boosterism was rife. League tables, sanctioned by the IOC in 1907 but revoked in 1914, became a central facet of the Games' fascination—the US press thundering its support for the home team in 1932, for instance, and rejoicing in its triumphs. Tensions surrounding football (soccer) friendlies between England and Italy/Germany in the mid 1930s, and the vituperative atmosphere in which the Max Schmeling–Joe Louis world heavyweight title bouts were fought in the same period point to sport's innate ability to transport political meanings.

Even before the Games began, the torch carried Germany in dramatic form across Eastern Europe too (from Olympia to Saloniki, Sofia, Belgrade, Budapest, Vienna, Prague, Dresden and Berlin), allowing its peoples to express their views of the regime (Bernett and Teichler, 1999). In retrospect, it has been claimed the relay (sponsored by the major armourer Krupp) prefigured Germany's bellicose intentions in the region, and although this smacks of lazy thinking, it is certainly important to register the various ways in which the flame was understood. Whilst generally greeted with euphoria— some stretches had to be run more than once, and tens of thousands gathered in cities even when it arrived in the dead of night—it was scarcely free of political nuance. In Athens the Horst-Wessel-Lied was slipped into the programme against original plans, and in Budapest the ceremony turned into a celebration of Greater Hungary. In Vienna Nazis read the flame as a triumphal German entry into the capital and, amidst chaotic scenes, took to the streets in their hundreds of thousands, protesting against the government and embroiling themselves in street-fighting. Czechoslovakia proved no less turbulent—left-wing Czechs protesting so vehemently that the runners had to be protected, and the German majority in the Sudetenland greeting it with outstretched arms and interpreted it as a call to return to the Reich. From the outset, therefore, the Games were inevitably entwined with the political climate of the day.

BOYCOTT DEBATES

It was in the West, however, that Germany's suitability as hosts had been most seriously debated after Hitler came to power and calls to boycott began to gather. The USA (as the leading Olympic nation), United Kingdom (the motherland of modern sport) and France (the founders of the modern Games) formed the crucial triumvirate whose absence would have severely damaged the Games, and the all-important Americans led the way. In 1933 IOC President Baillet-Latour requested that Germany produce a written guarantee of its commitment to the committee's regulations, with US member Charles Sherill further demanding that it reverse its decision to exclude Jewish athletes from its team. The pressure was relieved by a suitable statement from Berlin later that year, but the passing of the Nuremberg laws in 1935, which codified widespread discrimination against Jews, brought the issue to the boil again. In the heat of the debate, Avery Brundage (powerful President of the American Athletic Union) visited Germany, and Sherrill met with Hitler. Neither looked too closely: Brundage was given a less than candid account of conditions from Jewish sports functionaries due to the quietly menacing presence of German government officials in his entourage, and Hitler had to invite Sherrill back after he had stated with some petulance that German Jews would not be permitted to compete. In the USA Brundage used the popular Schmeling to vouch for his country's honesty, and Sherrill persuaded the Germans to include two half-Jews, living abroad, in their squad: fencer Helene Mayer and ice-hockey player Rudi Ball. The final vote, in December 1935, proved tight, the boycott being avoided by only 58.25 to 55.75. Two weeks before the Games, when it was too late for any action to be taken, the fully Jewish Gretel Bergmann was omitted from the high jump, despite equalling the German record that year.

In these episodes, individuals did not cover themselves in glory: Brundage's anti-Semitism would later become an open secret (although it did not stop him rising through the IOC's ranks to the Presidency), whilst Sherrill was a convinced anti-communist and admirer of Hitler and Italy's Benito Mussolini. Globally and locally, though, other considerations were at play, too. War-weariness and the fact that the boycott was supported predominantly by communists and trade unionists, who represented a greater threat to many than Nazism, tended to dampen enthusiasm for withdrawal. In 1935–36, the West did nothing to oppose Mussolini's invasion of Ethiopia and the remilitarization of the Rhineland, the violation of the Locarno pact. Nor did it intervene in the Spanish Civil War after Germany and Italy sent arms to

Francisco Franco. The USA was partially anesthetized by its own discrimination against African Americans and, pragmatically, its teams had been competing unhindered in Germany since 1933. For delicate domestic reasons, the US State Department and the President's Office remained determinedly neutral. In the United Kingdom, a growing mood of appeasement and the sports world's dogged insistence on its sacredly held belief in the separation of sport from politics put the team on the road to Berlin. In France, social divisions between bourgeois sport and its working-class counterpart—the former fully embracing Olympic ideals, the latter going its separate way and supporting the Workers' Olympics from 1925—caused the battle lines to be drawn up in such a way that even well-founded reservations on the right could not prevent participation. In France, too, Jewish support for a boycott was muted for fear of identification with the left, and in a spirit of uneasy compromise, the new socialist government under Léon Blum failed to resolve tensions by releasing funds for both Berlin and the rival socialist Games in Barcelona (which were eventually cancelled due to the Spanish Civil War) (Holt, 2003; Krüger, 2003a; Large, 2007; Murray, 2003b).

ORGANIZATION

The 1936 organizers and the regime profited from these dubious personal predilections, complex prevarications and mitigating circumstances. While the world debated, Germany acted, and did so with a singularity of purpose that had previously seemed doubtful. At the 1932 Olympics in Los Angeles, Lewald had told Baillet-Latour that Hitler would be opposed to staging the Games. Germany's hosting of the event was a specific remnant of the Weimar Republic's policy to reintegrate the defeated nation into the world community, and as vehicle for liberal, pacifist and international thinking, the Olympics were anathema to Nazi ideology. None the less, Hitler never missed an opportunity, and as with sport in general, he decided to embrace the Games for his own purposes. Several reasons will have informed his apparent volte-face. First, the Games would promote sport among German youth and build national strength. Second, success in the sporting realm would help the country regain its status as a world power. Third, the event itself would, as Hitler told Goebbels in 1933, 'impress world opinion by cultural means'. Finally, such cultural means would project a peaceful image whilst preparations of another kind continued apace. By October 1933, Germany was withdrawing from the League of Nations, and only two weeks after the Games, Goebbels unveiled a four-year plan that culminated in the nation readying itself for war by 1940.

From this cluster of motivations came Hitler's dictum that the Games should be 'the task of the whole nation', their preparations 'complete and magnificent'. As in Los Angeles four years earlier, employment was a key issue. 'When a nation has 4,000,000 unemployed', Hitler announced, 'it must seek ways and means of creating work for them'. His resonant municipal-economic war cry, 'We will build', led to the engagement of 500 firms on a daily basis for two and a half years. However, the fundamental difference was national backing. Although the State of California had issued bonds before the Wall Street collapse, the Federal Government restricted its financial input to some limited support for its athletes. Berlin, albeit for the government's own propaganda purposes, simply blurred the inherent contradiction in the Olympic Charter between the jealously guarded independence of the host city and organizing committee on the one hand and state approval (e.g. patronage) on the other. The actual unimportance of this distinction is evidenced in the fact that Diem and Lewald had to sign a statement promising they would maintain the existence of such independence internationally, whilst in fact the complete interpenetration of government and civil agencies in the organizing committee is openly recounted in the Official Report. Basing their plans on the abandoned 1916 blueprint, the organizers initially requested 6m. Reichsmark (RM, roughly £300,000) of governmental support, but received around RM 100m. instead from a regime intent on surpassing all that had gone before. Hitler's visual imagination was particularly fired by the possibilities of the Reichssportfeld, which he upgraded from the original idea merely to expand the 1916 stadium, transforming it into a spectacular site, complete with amphitheatre, commemorative hall, and abutting parade and display square for 250,000, enhanced by his own personal attention and finishing touches from Albert Speer.

Artistic vision and nationalistic branding were at the heart of the Olympic enterprise. De Coubertin had always thought of sport as part of high culture and envisaged the different components of the Games as a harmonious whole. Inspired both by Richard Wagner and the English philosopher John Ruskin, who created the British Arts and Crafts Movement, he conceived of the Games as not just containing art and music, but as forming a complete artwork in themselves, the sports event appearing 'in a unity of the athlete with the spectator, with the surroundings, the decoration, the landscape etc'. This underlying harmony was rarely recognized, never mind attempted, in Olympic cities. However, as the planning for Berlin as early as 1932 suggests, Diem, who was de Coubertin's primary exegete, had set himself the goal of turning the ideal into reality. Already in 1930 he had noted that 'the festive character of the former Olympic Games had left much to be desired from the viewpoint of harmony' and set about his principle strategy for achieving the 'Gesamtkunstwerk': the 'Olympische Jugend' (Olympic Youth), a Festival Play that eventually came to involve an overall cast of 10,000, singing and dancing to music by Carl Orff and Werner Egk, culminating in the choral movement of Beethoven's Ninth Symphony, which de Coubertin had been keen to introduce to Olympic ritual, and a 'Lichtdom', a cathedral of light the beams of which meet high above the stadium.

More strikingly, however, under the Nazis' guidance, the whole city turned itself into a 'Gesamtkunstwerk', treating itself to an Olympic make-over, both cosmetically and infrastructurally. 'Unsightly buildings and other "eyesores"', as the Official Report puts it, 'were covered with greenery', and considerable sums were poured into railway constructions around the Reichssportfeld and other transport projects around the city. The decorative scheme for the city was ordered by the Mayor and President of the Council in collaboration with the Reich Minister of Propaganda and 'combined utility with beauty, simplicity and colourfulness in a highly effective manner'. On the opening day, the Olympic event clicked smoothly into action across the capital: the Wehrmacht sounded a massive reveille, which was followed by children's sports displays in every part of the city and a huge procession down the Via Triumphalis to the stadium, where the opening ceremony was concluded after nightfall with Diem's play. Hundreds of thousands of people were in the right place at the right time. For this to be possible, every part of the metropolis, small and large, and every access route and entry point to it had to be in good working order. Graf Helldorf, President of the Berlin Police Force, banned the drying of washing on balconies and open windows during the Games, to preserve the pristine look of the facades. He also oversaw the remodelling of Berlin's road system, ridding important traffic-dense arteries of stationary vehicles and transforming less vital routes into complex parking lots. The population played its part—according to the Report, cheerfully, too. 'When the escorting police automobile announced the approach of an arriving team, traffic in all directions stopped automatically and the public formed welcoming lines on both sides of the streets.' The Games may be too casually remembered through the prism of Riefenstahl's epic but in August 1936, there was a cinematic feel about the city. 'It [was] no longer Berlin', as France's Le Jour put it, 'it [was] a film set' (Murray, 2003b).

REACTIONS

When assessing the impact of the Germans' efforts, it is important to distinguish between home and foreign audiences. Abroad, there was much praise for the high standards of organization and spectacle elements, but the boycott debates had raised awareness of the Games' façade, and most countries sent political as well as sports reporters who subjected the event to intense scrutiny. On the whole, the Games changed no one's opinion of the country, the pattern of coverage matching almost exactly views expressed in the contentious run-up. The regime had gone to great lengths to publicize the Games around the world that were hardly surpassed when the Federal Republic of Germany (West Germany) hosted the event in Munich in 1972. However, an internal report for the Ministry of Propaganda was forced to admit that attempts to influence overseas opinion had roundly failed. There were, of course, local variations in reporting. In the United Kingdom, Owens was admired as the supreme sportsman, although there was little sense of his performance having defied Nazi ideology. Overall, the British media were more concerned with consoling themselves for their own team's poor results, defending their athletes as 'true amateurs' in contrast to the professionally trained National Socialists. In France, too, the country's mediocre performance united the press in its condemnation of the government's indifference to sport and the National Olympic Committee's organizational incompetence. Their German neighbours came in for criticism, their dreadful food, lack of spontaneity and love of discipline being caricatured in various ways, and the two papers with the largest circulation (*L'Auto* and *Paris Soir*) joined their left-wing counterparts in lampooning Germany's overt nationalism, the former leading with the emotive headline 'J'accuse' and provoking a retort from de Coubertin himself. Only in the USA, where the boycott had been debated most acerbically and where Riefenstahl's 'Olympia' would meet with such uniform opposition on its release two years later, did coverage lack light and shade. The *New York Times* dubbed 1936 'the greatest propaganda stunt in history', another paper describing the Germans as 'past masters of kidding the public', who were 'now concentrating on kidding a few thousand athletes and officials' (Holt, 2003; Krüger, 2003b; Murray, 2003a, 2003b).

The happiness and good opinion of those sportsmen and women (4,066 of them from 49 nations, in fact, joining the 150,000 tourists) were important to the regime, but the effect of a fabulously organized and breathtaking event on its own public (some 3.5m. of them attending in person) was at least of equal significance. At any rate, it was in the domestic sphere that it registered its undoubted success. In late 1935, US diplomat George Messersmith had reported: 'The youth of Germany believe that National Socialist ideology is being rapidly accepted in other countries ... To the Party and to the youth of Germany, the holding of the Olympic Games in Berlin in 1936 has become the symbol of the conquest of the world by National Socialist doctrine. Should the Games not be held in Berlin, it would be one of the most serious blows which National Socialist prestige could suffer within an awakening Germany.' No such blow was delivered, however, and the German press could shield the public from the boycott discussion and relay only positive reports from abroad. Moreover, the team's performance underlined a sense of national buoyancy. The Games are remembered, rightly, for the achievements of the US black athletes (the 18 of them gathering half of the team's overall points), but the Olympics, as the Ministry of Sport's report concluded, belonged to Germany. The USA dominated track and field and swimming, but the German athletes—many of whom had been permitted, against Olympic regulations, to train on a professional basis—beat them into second place. Having dropped from second in Amsterdam 1928 to sixth in Los Angeles and enduring the Turners' derision of their meagre four gold medals, the Germans followed the example of fascist Italy (which had risen in the same period from fifth to second), by carrying off 33 gold, 26 silver and 30 bronze medals, with particularly strong showings in boxing and rowing. For a country still recovering from its long post-1918 humiliation, such results proved a major boost to national self-confidence. As the Party newspaper *Der Angriff* put it, it was 'truly difficult to bear so much joy'.

Whether the collective buzz had been generated by propaganda or, as claimed by one reputable historian (Eisenberg, 1999), the Germans threw themselves so completely into the Olympic spirit that their Nazi salutes actually became empty forms for two weeks, is impossible to say but there was certainly some truth behind the hyperbole. Sport's massive popularity—which had encouraged the Nazis, after all, to adopt it over Turnen—doubtless formed the bedrock of support and enthusiasm. One other historical certainty was Hitler's good standing with the German population, which peaked in 1936 (Grothe, 2008). In that year, the Winter and Summer Games bookended the re-occupation of the Rhineland and the successful national election, and the cult of the Führer reached its apotheosis. We must be cautious, however, not to overstate the case. On the one hand, a collapse of the Games might have dented incipient German hubris, as Messersmith dared to hope, but not such that war would have been avoided by the decade's end; on the other, the momentum gathering behind the regime and its leader would have done so, perhaps only marginally later, without the Olympics. The Games' contribution to the regime's popularity, however, cannot be gainsaid.

CONCLUSION

Whether the Games should have been boycotted or withdrawn from their contentious hosts by the IOC is a moot point. The reasons why nations and sports federations were hesitant at the time are fairly evident. After the fact, Jesse Owens argued the boycott would have deprived him of the chance to mock Hitler's Aryan policies—a view that would unite him in the 1960s with Avery Brundage, of all people, over radical debates in the era of black power struggle. Brundage, like de Coubertin before him, would cite the Movement's crucial mantra every time a boycott loomed (which it did again from the Suez crisis and the USSR's invasion of Hungary onwards): would the Games ever be held if peace and harmony were essential prerequisites? Leaving the parlour game of hindsight and retrospective ethics to one side, Berlin contains two important moments in Olympic history. First, it made the Games even more magnificent than Los Angeles through national, not just municipal, aspirations and finance. One might question the capillary menace with which these aspirations were sometimes achieved, but the consequences for the Olympics were clear: after Berlin, it would no longer be possible to host them without such backing. With the exception of the quirkily financed Los Angeles Games of 1984, every Olympics since—be it Beijing 2008 or London 2012—have been underwritten by governments, even in the age of commercialization. Second, the IOC maintained its momentum. As de Coubertin noted in an interview after the 1936 event, 'What's the difference between propaganda for tourism—like in the Los Angeles Olympics of 1932—or for a political regime? The most important thing is that the Olympic Movement made a successful step forward' (cited in Krüger, 2004: 37). Riefenstahl was awarded the Olympic Diploma in 1937, the Nazi leisure organization Kraft durch Freude the Pierre de Coubertin Cup in 1938, and Diem was allowed to establish an International Olympic Institute in Berlin in 1937. Travelling light, Olympism established itself as a creed for all-comers, a stance that would bring it into further stormy waters but ensure its ultimate survival. Berlin had been the test that strengthened the morally weak.

REFERENCES

Bernett, H. and H.J. Teichler (1999) 'Olympia unter dem Hakenkreuz', in M. Lämmer (ed.), *Deutschland in der Olympischen Bewegung: Eine Zwischenbilanz*. Frankfurt am Main: Nationales Olympisches Komitee für Deutschland, 127–71.

Eisenberg, C. (1999) *English Sports' und deutsche Bürger: Eine Gesellschaftsgeschichte 1800–1939*. Paderborn: Ferdinand Schöningh.

Graham, C.C. (1986) *Leni Riefenstahl and Olympia*. London: Scarecrow Press.

Grothe, E. (2008) 'Die Olympischen Spiele von 1936'. *Geschichte in Wissenschaft und Unterricht*, 59(5/6): 291–307.

Hoberman, J. (1995) 'Toward a Theory of Olympic Internationalism'. *Journal of Sport History*, 22: 1–37.

Holt, R. (2003) 'Great Britain: The Amateur Tradition', in A. Krüger and W. Murray (eds), *The Nazi Olympics: Sport, Politics and Appeasement in the 1930s*. Urbana and Chicago: University of Illinois Press, 70–86.

Keys, B.J. (2006) *Globalizing Sport: National Rivalry and International Community in the 1930s*. London: Harvard University Press.

Krüger, A. (2004) 'What's the Difference Between Propaganda for Tourism or for a Political Regime? Was the 1936 Olympics the First Postmodern Spectacle?', in J. Bale and M.K. Kristensen (eds), *Post-Olympism? Questioning Sport in the Twenty-First Century*. Oxford: Berg, 33–49.

——(2003a) 'Germany: The Propaganda Machine', in A. Krüger and W. Murray (eds), *The Nazi Olympics: Sport, Politics and Appeasement in the 1930s*. Urbana and Chicago: University of Illinois Press, 17–43.

——(2003b) 'United States of America: The Crucial Battle', in A. Krüger and W. Murray (eds), *The Nazi Olympics: Sport, Politics and Appeasement in the 1930s*. Urbana and Chicago: University of Illinois Press, 44–69.

Large, D.C. (2007) *Nazi Games: The Olympics of 1936*. New York: W.W. Norton.

Mandell, R.D. (1971) *The Nazi Olympics*. New York: Macmillan.

Murray, W. (2003a) 'Introduction', in A. Krüger and W. Murray (eds), *The Nazi Olympics: Sport, Politics and Appeasement in the 1930s*. Urbana and Chicago: University of Illinois Press, 1–15.

——(2003b) 'France: Liberty, Equality, and the Pursuit of Fraternity', in A. Krüger and W. Murray (eds), *The Nazi Olympics: Sport, Politics and Appeasement in the 1930s*. Urbana and Chicago: University of Illinois Press, 87–112.

Riordan, J. (1999) 'The Worker Sports Movement', in J. Riordan and A. Krüger (eds), *The International Politics of Sport in the Twentieth Century*. London: E&F.N. Spon, 105–20.

Roche, M. (2000) *Mega-Events and Modernity: Olympics and Expos in the Growth of Global Culture*. London: Routledge.

FURTHER READING

Bachrach, S.D. (2000) *Nazi Olympics: Berlin 1936*. USA: United States Holocaust Memorial Museum.

Mandell, R. (1987) *Nazi Olympics*. Champaign, IL: University of Illinois Press.

Walter, G. (2006) *Berlin Games: How the Nazis Stole the Olympic Games*. London: John Murray.

The Olympics and the Cold War: An Eastern European perspective

MIKLÓS HADAS

INTRODUCTION

When viewed as a microcosm of social relations, the Olympic Games provide a unique opportunity to examine the interplay of complex symbolic forces on the global scene. From the outset, international sport events have often been heavily influenced by conflicts, used and abused as they have been to represent a variety of political interests (Walters, 2006). The Olympic Games are a sporting mega-event par excellence and, as such, have been hijacked in various ways for political purposes. From an Eastern European perspective, this essay examines the intricacies of the Cold War and their affect on the Olympics.

The Cold War was a 40-year long politico-ideological conflict between two superpowers: the USSR and the USA (Dockrill, 1988). It represented a major clash over ideology, economic systems and global military presence, power and alliances. Although during and shortly after the second World War the USSR and the USA were friendly to one another, the Yalta conference (4–11 February 1945) saw the beginning of a deteriorating relationship. Driven by emerging tensions, leading to political anxieties, both of the Cold-Warring parties made attempts to strengthen their own political and military positions by increasing their weapons arsenals and recruiting allies (Dockrill, 1988). On the one hand, Joseph Stalin ensured that the countries of Eastern Europe were occupied by the Red Army and would become communist states. These included Lithuania, Estonia, Moldavia, Poland, Czechoslovakia, Hungary, Romania, East Germany, Yugoslavia and Bulgaria. The effect of this consolidation was to place an additional 100m. people under Soviet domination, Stalin's ultimate goal being the establishment of one-party communist states in the entire area surrounding the USSR. On the other hand, the USA helped to establish and support capitalist democracies in much of Western Europe, ensuring a decisive American military, economic and political role in the region and, thereby, securing a firm front-line against potential communist expansion.

The Olympic Games, between 1952 and 1988, can be seen as the extension of the Cold War via sporting symbolism: the 'capitalist' world being pitted against the 'socialist' world in the sporting arena (Andrews and Wagg, 2007). By operating a comprehensive, centrally controlled and heavily subsidized sports programme, representatives of some of the Eastern European socialist countries achieved outstanding Olympic and international sport records. Their achievements were the direct result of centrally-monitored sport systems allowing for the construction of sporting facilities, talent identification, coaching-related knowledge transfer, social rewards (both in the form of prestige and money) and, on occasion, centrally administered illegal performance enhancing practices (Dimeo, 2008). These symbolic (over-)investments were implemented to express politico-ideological superiority over the Western world, especially the USA.

The Soviet leadership prioritized establishing records during international competition. To this end they created an enormous sport bureaucracy that employed more than 280,000 people by 1976. In the Russian Republic alone there was a Central School for Trainers, seven institutes of physical culture and over 1,500 stadiums. The use of sport as a Cold War weapon reached its pinnacle after the Soviet invasion of Afghanistan in December 1979. US President Jimmy Carter announced his disapproval at the military manoeuvre and stated that the USA would boycott the 1980 Moscow Summer Games unless the USSR withdrew its military presence from the occupied territory. The USSR refused to comply and the USA, along with some 60 other nations, withdrew from the Games. This political gesture further deepened the ideological gap between the superpowers and precipitated the boycott of the 1984 Los Angles Summer Games by the USSR and most of its allies.

To understand the importance of international sporting achievements during the Cold War era, it is important to note that the USSR appeared on top of the unofficial medal table six times in this period (in 1956, 1960, 1972, 1976, 1980 and 1988). This is a reflection of colossal athletic achievement and financial and social investment, especially when one considers that the USSR had only joined the Olympic fraternity in 1951. The outstanding performance of the German Democratic Republic (East Germany) is also well known (in the unofficial medal table they were placed fifth in 1968, third in 1972 and second in 1976, 1980 and 1988), and can best be understood within the same context of political tension between East and West. Generally speaking, athletes of socialist countries exceeded their previous achievements due to the politically fuelled gold medal-rivalry between the countries and their ideological antagonists.

In this essay, the Olympic performances of the USSR and seven other Eastern European socialist countries (Bulgaria, Czechoslovakia, East Germany, Hungary, Poland, Romania and Yugoslavia) will be discussed, along with the underlying sport political strategies that facilitated these achievements. In doing so, attention will be given to political tensions and to symbolic distinctions and dispositions represented by sports.

INCREASING INTERNATIONAL SPORTING INTERESTS

The expansion of the Olympic Games is manifest in the Cold War period and can be illustrated with some descriptive data. For instance, whilst athletes from only 69 countries took part in the Helsinki Games in 1952, by the time of the Tokyo Olympics (1964), the number of participating countries had increased to 93. Although the international significance of the Olympics continued to ensure a steady increase, in 1976 the number of participating nations dropped slightly to 92 as a result of over 20 African states boycotting the Games in protest against the participation of New Zealand. This boycott, along with the ones four and eight years later, reduced the number of participating countries but also shed light on the politicization of the Olympic Games (www.kiat.net/olympics).

The continuing expansion and growing political significance of the Games prompted socialist countries to make increasingly greater efforts to be successful in order to showcase their ideological superiority. Thus, in 1952 the athletes of the Soviet bloc won 48 gold medals, one-third of all those available. In 1960 they won 61 gold medals, 40% of those on offer. In 1968 the gold medal count reached 69, increasing to 114 in 1976 and continuing to rise to 128 in 1988 in Seoul. It should be noted that at the Seoul Games, in which 159 countries were represented, the athletes of eight socialist countries won over 50% of all the gold medals.

Overall, the data show that during the Cold War, competitors from socialist countries won over 50% of the gold medals in nine different sports. There were specific sports in which socialist countries excelled throughout the Cold War, namely football (soccer; out of the eight gold medals, three were won by the Hungarians, two by the USSR and one each by the Yugoslavs, East Germans and Poles), water polo and handball. These results can be taken as indicative of how important team sports were for communist leaders, who viewed them as a vehicle for overtly propagandistic symbolic content relevant to the entire nation. In fact, it has been argued that team sports, especially football, were a state affair under communism (Hadas, 2000). By recycling national sporting traditions for propagandist purposes, the communist leadership could achieve multiple aims: they could offer positive identification patterns for their citizens, present sporting careers as an attractive opportunity for social mobility and legitimate the flexibility of the political system. Team sports have the tendency to express intra- and inter-community distinctions in a dramatic form (e.g. as a sequence of dramaturgical happenings), and hence they allow for many people to identify with the connotations implied and permanently redefined. In the early 20th century, these symbolic distinctions emerged mostly within the context of each sport's and each team's local and national embeddedness. In the beginning, they bore their national identity content spontaneously. From the Berlin Olympics onwards, however, a new tendency appeared: earlier spontaneous and often incidental connotations began to be put into the service of legitimating propaganda goals. Consequently, party leaders often utilized international team sport events to symbolize the struggle between communist and capitalist ideologies. In particular, the USSR made extraordinary efforts in this regard and, driven by their desire for medals, in addition to excellent results in handball, they won three gold medals in basketball and five in volleyball during the Cold War.

This same observation also applies to football, a sport that was highly popular not only in the countries of the former Soviet bloc but also in most of the rest of the world, including Western Europe. Football is capable of continuously reproducing the conditions on the basis of which the collective identity of large social groups can be defined. As a result, the fan is no longer a simple spectator but, as a virtual participant in the struggle, s/he identifies with the collective connotations implied by the team and thus becomes part of the symbolic fight, in this particular case, between East and West. In this situation, subconscious and emotional components of collective identity may burst to the surface. These components are retrieved mostly in the context of sports rivalry, owing to the presence of a significant other. Under the structural constraints of the symbolic fight, the elements of collective identity are redefined and reinforced; out of a multitude of features, with those that are most relevant in the context of a given opponent coming to the fore. The Olympics provide an excellent arena in this regard inasmuch as the dispositions of fans crystallized in the national past can be projected on to athletes representing another nation (instead of the customary adversaries at home) and the identity patterns formed within the domain of football become transferable to the context of other (not necessarily team) sports (see Hadas, 2000).

Other events were also monopolized by the Soviet bloc. Kayak-canoe meets were especially fruitful with competitors from five socialist countries winning nearly 81% of the gold medals in the Cold War Olympics (25 by the USSR, 10 by the East Germans, six by the Hungarians, six by the Romanians and one by the Czechoslovaks). Rowing proved to be another area of dominance, with socialist countries taking 56% of the gold medals (the East Germans winning 22 in four Olympics and the other seven countries achieving 17 in eight Olympic Games, 11 by the USSR, one by the Yugoslavs, two by the Czechoslovaks and two by the Bulgarians).

Another Soviet success story was in weightlifting where they seized two-thirds of all available gold medals. It is worthy of note that six gold medals were won by Soviet athletes in the heaviest weight category (in which no weightlifter of another socialist country managed to win). Thus the 'strongest man in the world' title was won by the USSR six times, with extensive symbolic significance in relation to the ongoing political arm wrestling between East and West. The gold medals won in weightlifting by the socialist countries were rather unevenly distributed (34 by USSR, seven by Bulgaria, four by Poland, one by Hungary, one by Czechoslovakia and one by East Germany).

Gymnastics could be considered one of the most successful sports for the Soviet bloc. This is indicated by the fact that in eight Games, 70% of all gold medals were won by gymnasts from five socialist countries. Soviet dominance was marked (with a total of 64 first places), but the number of medals (eight for Hungary, eight for Czechoslovakia, six for Romania, four for East Germany) indicated that other socialist countries also had world-famous gymnasts who shared in the success (for example, Czechoslovakia's Vèra Čáslavská won three gold medals in both 1964 and 1968, and Romanian

Nadia Comăneci took three gold medals in 1976). The value of these achievements was further enhanced by the fact that gymnastics is a prestigious sport having strong historic credentials as well as attracting considerable media attention.

Regarding combat sports, it was wrestling in which athletes of the socialist camp were most successful, winning 57% of the gold medals. The results were similar to those for gymnastics with most first places (50) being achieved by the USSR, with all the other socialist countries also contributing to the success (Bulgaria with 11, Hungary eight, Romania four, Czechoslovakia one, and the remainder, two gold medals). Thus, wrestling can rightly be seen as a sport representing the entire Soviet bloc. Similarly to weightlifting, most gold medallists (11) in the top categories was the USSR (the exception being Hungarians who also managed to take two gold medals in the heaviest category).

Although in fencing nearly two-thirds (58%) of gold medals were also won by contestants from the socialist bloc, caution must be taken before this result is interpreted as a success story for the whole bloc. One reason is that these medals were won by only four countries and another is that fencing was among those rare sports in which Hungary, and not the USSR, won most gold medals (16 and 15, respectively). In addition, Poland won four and Romania one. Furthermore, it was widely known in political and sports circles (especially in Hungary) that fencing was not only deeply rooted in the conservative-gentry national tradition, but had been particularly favoured by the rightist, 'counter-revolutionary' regime in the inter-war period. Hence, fencing was not fully considered as a communism-appropriate sport.

Boxing was another popular combat sport in the Soviet bloc, although socialist competitors 'only' won 41% of all gold medals (still an outstanding result as it was achieved collectively by eight countries out of 100 or more). What is more, boxers from each of the eight socialist countries managed to win gold medals—USSR (13), Poland (eight), Hungary (four), East Germany (four), Czechoslovakia (two), Bulgaria (two), and the rest (one each). The value of boxing gold medals was enhanced by the greater media attention given to the sport when compared with that afforded to wrestling and fencing. The performance of the socialist countries was also outstanding in pentathlon with over 70% of all gold medals won. However, what is true of fencing also applies to this sport. Athletes from four countries collected all of these medals, with Hungary ahead of the USSR again. Moreover, these first places were less evenly distributed than in fencing with half the available gold medals (eight) going to Hungary, three to the USSR and one to Poland.

On the basis of these statistics, it is possible to gain insight into the Soviet bloc's desire for international gold medals and into the qualities and dispositional patterns embodied by their athletes. The embodiment of a socialist type of human being appeared in the Olympics as co-operative and ambitious to win. S/he subordinated him- or herself to the goals of the team (that is, the political community) and was not only steadfast, persevering, disciplined, obedient, dexterous and tolerant of monotony, but was also capable of uncompromising struggle and endurance in the face of immense pressure.

INTRA-BLOC TENSIONS: 'BLOOD IN THE WATER'

A unique aspect of the Cold War was that the USSR had not only external tensions and oppositions to face, but also internal resistance from some of the annexed countries. The USSR's aggressive integration policy provoked nationalist sentiments and led the annexed nations to rebel. Given that one of Stalin's goals was to establish one-party communist states in the area surrounding the USSR, he wanted to abolish private economic enterprise, political debate and tolerance for cultural differences. The pursuit of these goals met with national resistance from the outset. Occasionally sharp intra-bloc oppositions manifested themselves during sports competition. Rinehart interprets this internal tension as:

> ... a middle area between outright, open rebellion and underground, passive resistance and it involves the socially-accepted practices of symbolically 'killing' one's enemy and regaining one's land while being monitored by the world according to the rules of sport.
>
> (Rinehart, 1996: 122)

A case in point was the famous Hungarian-Soviet water polo match in 1956, also referred to in the phrase, 'Blood in the Water' (Rinehart, 1996; 2007). This match and, indeed, the entire Melbourne Games were tainted not only with bloodshed in the swimming pool but also with blood simultaneously being spilled on the streets of Budapest.

During the communist era, Hungary was in a state of political dependency, economic interdependency and cultural suppression. It was Sovietized from the point of view of politics, economy and culture (Fehervary, 1990). At the Yalta Conference, the Western allies had granted the USSR the right to occupy the Eastern European states. The Yalta agreement allowed for the maintenance of an occupation force in Hungary only until the signing of a peace treaty. However, although this occupation was to be short, it did not make the lives of Hungarians easy. The major problem was that no agreement ever specified the size of the Soviet military presence that was to be allowed to remain in Hungary. Moreover, providing for the Soviet army was the responsibility of the Hungarian Government. In a defeated country, when feeding the indigenous population was creating serious enough difficulties, supplying provisions for such a large military force was no simple task (Fehervary, 1990). In addition, between the Yalta agreement and the peace treaty, more than 300,000 Hungarians were transported to Soviet labour camps in Siberia and almost 100,000 of these never saw Hungary again.

The much-awaited peace treaty signed outside Paris was directed by an Allied Control Commission (ACC) with a Soviet marshal at its head. After the peace treaty was finally signed, the Soviet troops gave no sign of future evacuation and it became obvious that Hungary had to prepare for an extended occupation. For how long no one could tell (Fehervary, 1990). By this time the size of the Soviet army had been reduced by 500,000 soldiers and the troops who remained as a part of standing occupation, numbering over 100,000, were accommodated in military barracks.

In 1945, at hastily held elections, the communists gained only 17% of the votes (this was the beginning of communism in Hungary), but the leader of the ACC ordered the Smallholders' Party, which won the election, to form a coalition with the Communist Party. The last relatively free election, in 1947, was followed by years of communist control: show trials, executions, forced settlement of hundreds of thousands of people, imprisonment, harassment, forced industrial development, a decline in living standards and Stalinist dictatorship. The first step of the Communist Party was to eliminate its political opponents. The AVH (the State Defence Office) arrested leading members of the Hungarian political community, mainly members of the Smallholders' Party, and charged them with conspiracy against the new coalition-led Republic. After undermining the Smallholders by arresting their leaders, the communists forced the Social Democratic Party into a merger and, in 1948, the victory of 'the dictatorship of the proletariat was complete' (Fehervary, 1990). In the same year, the government started to seal the borders of the country and a few years after the 'liberation' of Hungary by the Soviet troops, the entire population was effectively imprisoned (Fehervary, 1990).

In addition, the structure of society was transformed. The government, according to one of the main ideas of communism, tried to change the multi-class society into a one-class (working class) society. The leaders of the Communist Party, Mátyás Rákosi and his men, did not understand the nation's character and subordinated Hungarian interests to those of the USSR. Instead of developing their own socialist society, they brought Stalin's system of terror to the country. During the years between 1949 and 1956, more than 700,000 citizens were arrested for political reasons in addition to the approximately 200,000 criminal cases (Fehervary, 1990).

Although the Communist Party ruled the country, most Hungarians remained anti-USSR. Therefore, the leaders of the Party decided to launch an extensive re-education process aimed at convincing people to accept not only communist ideology but also the idea of belonging to the USSR. The first target for this propaganda were the schools and the education system in general. It became compulsory for educators to enrol in various Marxist courses. Everything related to Western culture was to be denigrated, distorted or denied. The greatest damage was done via the falsification of history. Texts were rewritten to meet the dictates of Marxism, and everything noble or glorious about Hungarian history was either erased or reinterpreted according to communist doctrines. Students were expected to know the details of the lives of Vladimir Lenin and Stalin rather than the names and achievements of Hungarian historical figures (Fehervary, 1990).

The Communist Party sought an anti-religious, workers' controlled society. Inevitably this met with massive resistance in Hungary, where the Roman Catholic Church expressed its opposition to communist repression and was consequently persecuted. Some outstanding leaders of the Catholic Church succeeded in weakening the authority of the Communist Party, but were captured and, based on falsified accusations, were sentenced to prison. In reaction, a revolution broke out against Stalinism on the 23 October 1956. It was mainly the younger generation who took part in this freedom fight, angered as they were by the humiliating Stalinist rule and ritual discipline. On 23 October student demonstrations in downtown Budapest and the unauthorized shooting of demonstrators led to chaos. On the same night, Imre Nagy was appointed prime minister, a position he held for little more than 10 days. One of the main goals of this freedom fight was to remove the yoke of the USSR. Because it was the common desire of the Hungarian citizens, Nagy decided to leave the Warsaw Pact and declared neutrality. It appeared that the USSR was ready and willing to negotiate about withdrawing its troops. On 3 November, however, a Hungarian delegation that was supposed to meet with the Soviet army general was ambushed and arrested at the Soviet military headquarters. One day later, the Soviet army started an attack on Budapest and within a few days had destroyed the resistance movement.

Against this politically charged backdrop, the ill-fated October revolution against the USSR resulting in the deaths of thousands of Hungarians and the subsequent voluntary exile of 200,000 more, the Hungarian water polo team faced the USSR in the sporting arena at the Melbourne Games. Water polo matches between the USSR and Hungary were usually intensely emotional but no game in the history of the sport could equal this one for unrestrained ferocity. The two teams grappled from the opening whistle, both under and above the water, as hundreds of Hungarian expatriates in the crowd of 5,500 jeered the USSR, and waved the flag of freedom adopted by the revolutionaries. The avenging Hungarians did all the scoring, but the match grew ugly when, with a minute left to play and Hungary leading 4–0, Hungarian player Ervin Zádor was sucker-punched resulting in a wound that required 13 stitches. Police had to step in to prevent a riot, and the game ended with Hungary winning 4–0. A wire-service photograph of Zádor, standing on the pool deck in his trunks, dazed, blood streaming down his face, was published in newspapers and magazines around the world.

Exhilarated by their victory over the USSR, the Hungarians went on to beat Yugoslavia 2–1 for their fourth Olympic water polo gold medal and, as the team mounted the victory stand for the medal ceremony, Zádor burst into tears. 'I was crying for Hungary', he said, 'because I knew I wouldn't be returning home' (see iconicphotos.wordpress.com/category/sports). Indeed, fully half of the 100-member Hungarian Olympic delegation, including Ervin Zádor, defected after the Melbourne Games. Rinehart (1996: 133) provides the following synopsis of these events:

The 1956 Melbourne Olympics created high visibility and opportunity for Hungarian athletes, which meant that their participation and subsequent victories could form oppositional signs of political resistance to the USSR regime in Hungary. The athletes believed they could demonstrate Hungary's resilience to the world. Additionally, expatriates exploited every symbolic opportunity ... to show the world Hungarian tenacity and pride.

POLITICAL SUBTLETIES OF THE OLYMPICS

If the subtle political implications of the Olympics are to be further analysed, communist women's role in the Games should also to be considered. The simplistic dichotomy of 'fighting man versus pleasing woman' is refuted on the basis of the Olympic programme, since women not only compete and fight but, in some sports (synchronized diving and gymnastics, for example), men also represent beauty, the masculine body becoming an object of erotic desire. However, it is also true that the 'Olympic woman' is still required to fight in gentler forms than men. Moreover, the dual combat sports, based on the civilized drives of the *libido dominandi* and the individual sports competitions of disciplined, purposeful actors under standardized circumstances still mainly belong to the masculine realm. According to Bourdieu (2001), *libido dominandi* is a desire after domination, a sort of sense of duty based on an inner drive that a man 'owes himself', acquired unconsciously in the course of socialization. This drive, or *illusion dominandi*, is constitutive of masculinity, and causes men to be socially instituted to let themselves be caught up, like children, in the games of domination that are socially assigned to them, of which war is the ultimate expression.

The sportswomen of the eight socialist countries won 57% of all gold medals in eight Olympic Games, an even greater contribution to the success of the Soviet bloc than that of the men. Their achievement in 1976—winning 90% of the gold medals—was unparalleled. The East Germans made the major contribution to this success, getting 25 out of the 44 gold medals won by socialist women athletes in Montréal (the East German Olympic team, placed second in the unofficial table behind the USSR, took home a total of 90 medals, 40 of them gold). In swimming events for women, they won 11 of the 13 available gold medals. Female athletes from the eight countries achieved most in gymnastics, winning 50 of the available 51 gold medals in the period. This medal haul was distributed among five countries—the USSR (30), Czechoslovakia (seven), Romania (six), Hungary (five) and East Germany (two). Only at the Games in Helsinki in 1952 did a non-socialist sportswoman (a Swede) win a gold medal in gymnastics. Eastern European sportswomen also proved extraordinarily effective in kayak-canoe and rowing. These two sports, however, remained the monopoly of two countries—the USSR and East Germany—throughout. In athletics and swimming (in which the socialist women won 59% and 32% of gold medals, respectively), the performance of the eight countries fluctuated, although the East Germans were largely responsible for improved performances from the 1970s onward.

Breaking down to smaller components the results in athletics, one finds an interesting relationship. In individual athletic competitions all 24 of the gold medals awarded in the eight Olympic Games were taken by Eastern European female athletes. In the individual throwing competitions (shot, javelin and hammer) sportswomen of six countries shared the gold medals, with the USSR taking 12, East Germany five, Romania two, Czechoslovakia two, and Bulgaria and Hungary one each.

CONCLUSION

In sum, it can be stated that, mainly by appearing in the Olympic Games as individuals, the socialist athletes were saliently effective in a range of sports. The socialist women and men showed themselves to be persevering, disciplined, obedient and dexterous human beings not only tolerating monotony and desiring to please in a discrete manner, but also capable of enduring immense pressure. To represent the countries of the communist bloc was a sensitive issue because athletes symbolized not only their own countries but communism as well. The communist bloc continually wanted to demonstrate the superiority of their athletes, the very products of communist ideology, over those of 'frivolous' capitalist societies through grandiose sport achievements (Molnar, 2007). However, these forced attempts to demonstrate superiority and aggressive integration policies provoked both internal and external political and financial pressures, contributing to the ultimate demise of the 'strong bastion' of communism in the east (Molnar, 2007).

If sporting achievements are to be conceived as indicative of a country's sports policy and investment, the trends discussed above might also be interpreted as signs of latent symbolic internal conflicts. This refers not only to such spectacular sources of conflict as the participation of the Romanians at the Los Angeles Games as the only socialist country to take part but also to the differences between the achievements of certain members of the bloc demonstrable in their Olympic performance. Although outstanding athletes and their coaches achieved high status in the countries of the Eastern bloc, many of them defected to seek freedom and/or lucrative sporting opportunities. A few gymnastics-related examples of this migratory trend are the Russian-born Olga Korbut and the Romanian Nadia Comăneci and her coach Béla Károlyi. The pressure on elite athletes to perform caused great physical and emotional stress to communist athletes, causing illness and even permanent disability for some. Following the collapse of the USSR, both Olympic weightlifter Yuri Vlasov and chess master Gary Kasparov went as far as attacking the old USSR sports system for its exploitation of athletes.

As stated at the outset, through comprehensive, centrally controlled and heavily subsidized sports programmes, representatives of most of the Eastern European socialist countries achieved outstanding sporting success particularly in successive Olympic Games. These achievements came at a cost that was borne not least by many of the athletes themselves. As an exercise in waging war without weapons, however, the Cold War sporting rivalry and also those internecine disputes conducted on the track and in the pool, the Soviet bloc's strategy provides an almost unparalleled example of the interplay between sport and politics.

REFERENCES

Andrews, D.L. and S. Wagg (2007) 'Introduction: War Minus the Shooting?', in S. Wagg and D.L. Andrews (eds), *East Plays West: Sport and the Cold War*. London: Routledge, 1–9.

Bourdieu, P. (2001) *Masculine Domination*. Cambridge: Polity Press.

Dimeo, P. (2008) 'Good versus evil? Drugs, sport and the Cold War', in S. Wagg and D.L. Andrews (eds), *East Plays West: Sport and the Cold War*. London: Routledge, 149–62.

Dockrill, M. (1988) *The Cold War, 1945–1963*. Houndmills, Basingstoke: Macmillan.

Fehervary, I. (1990) *The Long Road to Revolution: The Hungarian Gulag, 1945–1956*. Santa Fe: Pro Libertate Publishing.

Hadas, M. (2000) 'Football and Social Identity. The Case of Hungary in the 20th Century'. *The Sports Historian* 20(2): 43–66.

——(2007) 'Gymnastic Exercises, or "work wrapped in the gown of youthful joy": Masculinities and the Civilizing Process in 19th century Hungary'. *Journal of Social History* 41(1): 161–80.

Molnar, G. (2007) 'Hungarian Football: A Socio-historical Overview'. *Sport in History* 27(2): 293–317.

Rinehart, R.E. (1996) '"Fist flew and blood flowed": Symbolic Resistance and International Response in Hungarian Water Polo at the Melbourne Olympics, 1956'. *Journal of Sport History* 23(2): 120–39.

——(2007) 'Cold War Expatriate Sport: Symbolic Resistance and International Response in Hungarian Water Polo at the Melbourne Olympics, 1956', in S. Wagg and D.L. Andrews (eds), *East Plays West: Sport and the Cold War*. London: Routledge, 45–63.

Walters, G. (2006) *Berlin Games: How Hitler stole the Olympics*. London: John Murray.

www.kiat.net/olympics (accessed 15 December 2008).

www.olympic.org/uk/organisation/missions/women/index_uk.asp (accessed 15 December 2008).

FURTHER READINGS

Caute, D. (2003) *The Dancer Defects: The Struggle for Cultural Supremacy during the Cold War*. Oxford: Oxford University Press.

Maraniss, D. (2009) *Rome 1960: The Olympics That Changed the World*. London: Simon & Schuster.

Walker, M. (1994) *The Cold War: A History*. New York: Henry Holt.

The Olympics in the post-Soviet era: The case of the two Koreas

JUNG WOO LEE

INTRODUCTION

The Korean peninsula is the last remnant of the Cold War. Ideologically, the two Koreas still maintain different political systems: communist North and capitalist South. While this ideological difference poses potential political problems, the struggle largely revolves around the question of reunification (Cummings, 2005). This issue has concerned the two Korean governments in the post-Soviet era. Ever since the division, reunification has been a central issue in Korean political discourse. No consensus has ever been reached as the two Koreas retain mutually incompatible ideas regarding reunification, even though both countries generally agree they should be integrated.

After the First Korean Summit in 2000, inter-Korean relations changed. In their Joint Declaration, the leaders of the two Korean states agreed that there was in fact no conflicting element in their approaches to reunification, and that they would make a concerted effort to realize this goal. As part of the effort, the two also consented to develop economic and cultural ties. Therefore, numerous inter-Korean co-operative exchange projects were initiated, including sport programmes. It seemed that the Democratic People's Republic of Korea (North Korea) and the Republic of Korea (South Korea) had taken a significant step toward reunification.

However, with a new conservative president in power in South Korea from early 2008, inter-Korean relations have become stagnant. Claiming that the previous government's humanitarianism toward North Korea helped that country develop nuclear weapons, the current South Korean conservative regime says it will take a tougher attitude towards its northern neighbour. This statement resulted in North Korea halting many co-operative cultural and economic initiatives developed over the past decade. Consequently a cold atmosphere again hangs over the Korean peninsula.

Due to the unpredictable, unstable relations between North and South Korea, many Koreans are ambivalent when delegates of the two Koreas encounter each other at Olympic venues; hostility and fraternity overlap depending on the particular political circumstances of the day. All of this provides interesting material that can be used to determine the political value

117

of the Olympics for the two states. This chapter discusses inter-Korean relations within the context of the Olympic Games from 1988 to the present.

THE SEOUL OLYMPIC GAMES

The 1988 Seoul Olympic Games were regarded as the last Cold War Olympics. After the two Olympic boycotts in 1980 and 1984, South Korea seemed to be a place where athletes from both capitalist and communist states could freely participate in sports without serious political problems (Senn, 1999). It seemed that the Seoul Olympic Games signalled the beginning of the end of the Cold War, which would actually occur a few years later. A paradoxical feature was that while South Korea successfully delivered a message of détente to the world, the country itself failed to resolve its ideological conflict with the North.

In fact, the Seoul Olympic Games increased the tension between North and South Korea to some extent (Larson and Park, 1993). It should be noted that the two Koreas were not fully recognized by the United Nations when the Olympic Games were staged in the South Korean capital. A serious dispute between the two Koreas over the legitimacy of their respective governments prevented them from being member states of the United Nations. South Korea castigated the North for being communist, stating that communism was alien to the entire Korean history, and North Korea criticized the South for being a puppet of American imperialists. Under these circumstances, hosting the Olympic Games could offer an advantageous position to declare political legitimacy for the relevant government to the rest of the world. Understandably, the Seoul Games offered the South Korean regime an opportunity to justify its political authority within the nation. South Korea's rapid economic development also increased its political weight in the region. For North Koreans, however, the Games to be held in South Korea were perceived as a crisis that could undermine the authority of the communist government. Therefore, the International Olympic Committee's (IOC) decision to award the 24th Summer Olympic Games to Seoul put the North Korean Government in a politically uncomfortable situation. Reflecting this, the North Korean Workers' Party newspaper wrote that 'never in the history of the Olympic Movement spanning nearly 100 years have the Olympic Games been hosted by a puppet in a colony' (*Rodong Shinmun*, 1985). In this context, it was clear that the Seoul Games would inevitably intensify tensions between North and South Korea.

Recognizing that the Seoul Olympic Games would bring undesirable consequences, the North Korean Government attempted to prevent its southern neighbour from hosting the Games. North Korea presented a written petition to the IOC and other international sport governing bodies indicating that South Korea was so politically unstable that the history of the Olympic Games would be seriously tainted if the sporting event proceeded in Seoul. The North Korean criticism of South Korean politics was not totally groundless. In fact, pro-democracy political demonstrations against the then military regime took place throughout the 1980s in South Korea. Large numbers of students, intellectuals and workers were involved, and the government deployed the police to repress the protestors. Until 1987, when the ruling party guaranteed that a democratic presidential election would be held at the end of the year, large numbers of people continually gathered in Seoul and other major South Korean cities to protest against the repressive military government. Some South Koreans also opposed holding the Seoul Olympic Games, stressing that the event would simply justify the existence of the military dictatorship. They further argued that the Seoul Games would solidify the division between North and South Korea. In addition, some protesters argued that the Olympic Games would only benefit a dominant class at the expense of the people's interests. This severe nation-wide clash between the government and the people gave North Korea good cause to appeal to the world to nullify the IOC's decision to allow the Olympics to take place in Seoul. Many countries agreed that the Olympics should be relocated, but the IOC stood its ground, stating that the Games must take place as scheduled.

The communist North also doubted South Korea's capacity to provide officials and athletes with a secure environment so they could concentrate on performance without worrying about safety. Indeed, in an attempt to disrupt South Korea's efforts to host the Olympics, in 1986 North Korea planted a bomb at Kimpo airport, the major international airport near Seoul; five people were killed and more than 30 were injured. The attack damaged South Korea's reputation to a degree, as the incident happened just before the opening of the Asian Games that were also held in Seoul. In addition, North Korean secret agents attacked a Korean airplane in 1987, killing all 115 passengers, in an attempt to destabilize South Korean society before the Olympic Games. The two terrorist acts shocked both South Koreans and foreign nationals alike. Such inhumane and immoral acts against South Korea caused the world to question North Korea's criticism of South Korea. Thus, while the communist regime succeeded in attacking the South before the Olympics, it failed to persuade the international community to support its position.

In addition to the criticism and the terrorist attacks, the communist regime also proposed jointly hosting the Olympics to prevent the Games from taking place in South Korea alone. The communist North Korea suggested that the two Koreas co-host the event and officially call the Games the 'Korea Olympic Games' or the 'Korea Pyongyang-Seoul Olympic Games'. The North also proposed a joint organizing committee to consist of two chairmen, one from North Korea and one from South Korea. North Korea also argued that the disciplines of the Games should be equally allotted to Seoul and Pyongyang and that the sporting union between North and South Korea should be arranged to allow for a single Korean delegation to participate in the Olympics. Some South Koreans, particularly those who protested against the military regime, welcomed the idea of co-hosting the Games. Nevertheless, having already formed an organizing committee for the Seoul Olympics, and having undertaken the necessary construction of facilities, South Korea could not accept the North's proposal in full. Instead, the South conceded that some events, such as football, might be held in North Korea if the communist regime supported the Seoul Games. Despite this offer, North Korea stubbornly insisted that its original proposal be adopted without change, and such an inflexible stance made discussion difficult. In the end, the debate generated no results, and North Korea finally declared an Olympic boycott in January 1988.

The Olympic Charter states that international friendship and fraternity are essential values of the Olympic Movement. The Seoul Games fulfilled these ideals to some extent by inviting countries from both the Western and Eastern blocs. No serious political issues based on ideological conflicts arose during the event. Thus, it was ironic that, whether intended or not, the Seoul Olympic Games deepened the conflict between North and South Korea. The South attempted to use the Olympics to display its power to the world, thereby elevating its status by comparison with the North. North Korea feared this and tried to disrupt the Seoul Games. Naturally, these rival positions increased tensions between the two Koreas. Put simply, whilst the Olympic Games enhanced international co-operation, it negatively impacted on inter-Korean relations.

SPORT AND THE POST-SEOUL GAMES POLITICAL STRUCTURE

Although North Korea maintained its hostility towards South Korea in the period immediately after the Seoul Olympic Games, rapid political changes throughout the Korean peninsula soon influenced inter-Korean relations. While the roots of the ideological conflict between North and South Korea still remain unresolved, the breakdown of Soviet-led communism significantly impacted on inter-Korean relations. Amongst other things, the most distinctive change that the demise of the USSR brought to the political landscape of the Korean peninsula was the development of successful diplomatic ties between South Korea and former communist countries in Europe. During the Cold War, no communist state had an official relationship with capitalist South Korea. In fact, the ideologically divided Korean peninsula was viewed as a microcosm of the political competition between the USA and the USSR, and understandably the South did not formally recognize any communist regimes. However, a mood of détente in US–Soviet relations was emerging in the late 1980s. Reflecting this, the South Korean Government implemented *Nordpolitik*, a foreign policy that attempted to build diplomatic ties with the People's Republic of China, the USSR and North Korea. In so doing, South Korea tried to weaken the Korean peninsula's conflict-laden political structure with the ultimate aim of possible reunification. Coincidently, the Olympic Games had taken place in South Korea in the midst of the US–Soviet reconciliation, and the occasion had positively influenced South Korean policy towards communist states.

Whilst preparing for the Olympic Games, South Korea had invited representatives of every IOC member state, including communist ones, to Seoul to help to promote the sporting event as 'the best ever'. This was the rationale for attempting to establish official relations with communist states. The USSR, Eastern European states and China responded positively to South Korean efforts; only North Korea refused and finally boycotted the event. Although the South failed to improve its relations with the North, the *Nordpolitik* overtures were largely successful as most communist countries now recognized South Korea as an official diplomatic partner. Without doubt, the Seoul Olympic Games was a catalyst in establishing official ties between South Korea and these communist states, and the volume of cultural and economic exchanges between them significantly increased after 1988. In other words, the South Korean Government to some extent used the Olympics to open diplomatic relations with communist countries.

Even though North Korea initially rejected South Korea's political gestures and condemned it for hosting the Games alone, South Korea's strategy of forging formal ties with North Korea's closest allies, the USSR and China, challenged the communist North's political identity. Moreover, although the North Korean political system did not solely depend on European communist regimes, perestroika (reconstruction) and the subsequent breakdown of the USSR shocked the North Korean political elite. These developments required North Korea to modify its attitude towards South Korea. If North Korea continued its hostilities toward South Korea, it would become politically isolated in the region. Therefore, while North Korea maintained its communist political structure without help from its traditional allies, its policy towards the South had to be rethought.

Inter-Korean sporting relations reflected this. Since the early 1990s, bilateral discussions about sports between the two Koreas have generated some meaningful outcomes. In 1989 North Korea proposed an inter-Korean dialogue to discuss creating a unified Korean team for the 1990 Beijing Asian Games. Although ultimately a united team was not sent to Beijing, some significant agreements were reached. First, the two Koreas agreed to use the Korea Peninsula flag instead of two separate national flags whenever participating in international sporting events as a unified team. Second, the two countries chose the traditional Korean folk song *Arirang* as the official anthem of the united Korean team. Third, they decided to use 'Korea' as the official name when a united team participated in international sporting competitions. Before the Asian Games, these three things had been major obstacles preventing the two sides from creating unified teams for international sporting events. Given that flags, anthems and names are closely related to national identities, the difficulty in reaching the agreement was understandable whilst the two countries were in such serious conflict. With this politically sensitive agreement, they overcame some of the toughest barriers hindering the smooth development of inter-Korean sporting relations.

The behaviour of the two Korean teams and their fans also hinted at improved relations between North and South Korea. Unlike at other international sporting competitions in the past at which the communist and capitalist Koreas were explicitly hostile, they were friendlier at the Beijing Asian Games. The athletes and officials greeted and encouraged each other when they met at the venues. Moreover, North and South Korean fans intermingled and supported all Korean athletes without distinction. They also waved Korean Peninsula flags and sang *Arirang* at the stadium. It was the first time that North and South Koreans had congregated in such an amicable way. Thus, even though the two Koreas took part in the Asian Games separately, the attitudes of Koreans at the Games demonstrated a thawing in North–South relations.

This mood continued in 1991. North and South Korea successfully created unified Korean teams in football and table tennis. The united Korean youth football team was formed in preparation for the FIFA World Youth Championship. In this competition, the Korean team advanced to the quarter finals, a moderate success in the context of Korean football history. In addition, the women's doubles pairing representing the unified table tennis team at the World Table Tennis Championship won that competition. These two successful sporting events generated the idea that the power of the nation would be maximized if North and South Korea co-operated. Inter-Korean sporting collaboration implied that the two Koreas were gradually moving towards reunification in the early 1990s. No one could imagine the political turmoil about to erupt.

REVIVED NORTH KOREAN COMMUNISM

North Korea effectively revamped its communist system so that the government could sustain its socialist political structure. North Korean developed 'our style socialism', which stressed that the country must keep its own socialist values in contrast to the European and Chinese communist states that had adopted capitalist economic systems. A most distinctive feature of 'our style socialism' is its nationalistic character (Chen and Lee, 2007). So nationalistic was North Korea's political ideology that the country became more antagonistic toward nations that challenged its political legitimacy. North Korea was especially hostile towards the USA because the latter maintained an intimidating anti-communist policy towards the whole North-East Asian region, believing this would accelerate the collapse of the North Korean communist regime. North Korea responded by maintaining a hardline communist policy based on 'our style socialism', describing the Americans as imperialists who were attempting to infringe on North Korean sovereignty. Consequently, North Korea declared its intention to withdraw from the Nuclear Non-Proliferation Treaty (NPT) in 1993. Both the USA's intimidation and North Korea's radical reaction increased tensions on the Korean peninsula.

Thorny US–North Korean relations resulted in a deterioration of inter-Korean relations, as the USA was South Korea's closest ally. Inter-Korean sporting relations reflected the political situation. After the Beijing Asian Games and the successful performances of the unified football and table tennis teams, many Koreans expected even closer sporting union at the 1992 Barcelona Olympics. However, political relations between the two Koreas began to freeze again, and no such union emerged. Owing to unfolding political problems in the region, North Korea coldly rejected the South Korean proposal to hold bilateral discussions for creating a unified Korean Olympic team. Instead, the North sent its own Olympic team to Barcelona and Atlanta in 1992 and 1996, respectively, dismissing the possibility of a unified Korean team. This bitterness continued until the end of the 1990s.

It should be noted that North Korea spent considerable resources on elite sports, and as a result, the country's sporting capacity increased remarkably in the 1990s. This development seems closely related to its nationalistic socialist policy and the unexpected financial problems that the state was experiencing (Lee and Bairner, 2009). The communist government needed a vehicle for propaganda that motivated its people to partake in the economic and ideological 'Ardent March' campaign that was designed to overcome financial and political problems. Aware that Olympic success could be politically exploited to help promote reunification, the communist government strategically fostered athletes whose sporting talents were competitive on an international level and who had the potential to win a medal for their home nation. North Korea sent a relatively large Olympic team to Barcelona, and won nine medals including four golds. North Korea also participated in the Atlanta Games, winning five medals. The state media eloquently lauded the medal winners, and called them communist role models who achieved great things for the fatherland despite the lack of resources. The political use of sports, and of the Olympic medallists in particular, to sustain a social order by the North Korean Government became evident with its nationalistic socialist policy.

SPORTING UNION

Inter-Korean relations began to thaw again in the late 1990s. The South Korean regime elected in 1997 adopted a soft approach toward its Northern neighbour, called the 'Sunshine Policy' (Son, 2006). This policy was based on coexistence and constructive engagement with North Korea. The government also actively pursued dialogue and co-operation, and sought a gradual reconciliation between the two Koreas. The Inter-Korean Summit took place in the North Korean capital of Pyongyang for the first time. As a result, the overall Korean political picture suggested improved relations between the North and South, although conflict still existed around the armistice line dividing the two.

The 'Sunshine Policy' also promoted an environment in which an inter-Korean sporting relationship thrived (Merkel, 2008). A number of sporting exchange programmes were carried out, and some South Korean companies such as Hyundai gave the North technical and financial support to build a new stadium in Pyongyang. The most dramatic sporting event demonstrating the reconciliation of the two Koreas was the joint march at the opening ceremony of the Sydney Summer Olympic Games in 2000. It was the first time that the North and South Korean teams had entered an Olympic stadium together, bearing the Korean Peninsula flag and wearing identical outfits. Although the two Koreas took part in the Olympics separately, this event was sufficient to show the effect of the 'Sunshine Policy' on sport. Not only did the joint march symbolize a reunification of the two states, but it also displayed a unified Korean national identity to the world. Since then, the joint march has been the norm when the two Koreas have participated in any international competition.

Improved inter-Korean relations were also evident at the 14th Asian Games which took place in 2002 in Busan, the largest port city in South Korea. Unlike on past occasions when North Korea had boycotted any international sporting competition held in South Korea, this time the communist North sent an unprecedented delegation to the capitalist South. More than 600 North Korean nationals, including staff members and supporters, crossed the border, the first time that such a large number of North Korean individuals had made an official visit to South Korea since the peninsula's division. Most South Korean people welcomed the visitors from the North, and the athletes from the two Koreas encouraged each other whenever they encountered. The two Korean teams displayed such a friendly and mutually supportive attitude that the leaders of the two national teams praised their conduct. For instance, at the end of the Game Ung Chang, a North Korean member of the IOC, remarked that he was very much pleased to see that the two Korean states, which are in fact a single ethnic nation, co-operated well at the competition in spite of the fact that the North Korean side had not won as many medals as it had expected. Yun-taek Lee, the President of the South Korean national Olympic committee (NOC), shared this sentiment stating that the Busan Asian Games would, without doubt, facilitate inter-Korean sporting exchanges and co-operation programmes. In addition, some senior members of the two Korean teams discussed a possible sporting union for future Asian and Olympic Games, and all agreed that North and South Korea would make their best efforts to forge a unified Korean team for future international sporting competitions. Overall, the North and South Korean people shared a festive atmosphere largely derived from the improved relations between the two Korean states during the Asian Games.

While a united Korean Olympic team still remained to be seen, co-operative sporting relations continued at the 2004 Athens Olympic Games (Lee and Maguire, 2009a). There were numerous occasions when the members of the two Korean Olympic teams met and were photographed together. The leaders of the two Korean NOCs met officially with the President of the IOC to discuss possible sporting union at future Olympic Games. As before, the two

Korean teams marched together at the opening ceremony. However, the most remarkable event was the joint training session between the North and South Korean table tennis teams. While this was only a one-off event lasting just over an hour, the political implications of the session could not be ignored. The IOC welcomed the event on the grounds that it could contribute to realizing Olympism at the Games, and the NOCs of North and South Korea claimed this event could help in creating a united Korean Olympic team in the future. Moreover, when the members of the two table tennis teams met, they displayed a sort of political gesture, signifying the emotional meeting of the two Koreas, through actions such as hugging each other—an overt expression of emotion in the context of Korean culture. The media also praised this event. For instance, South Korean media portrayals of the event were replete with images and texts highlighting the symbolic reconciliation and reunification of North and South Korea (Lee and Maguire, 2009b). In particular, the media attempted to reinforce the notion of the inter-Korean sporting union by relating the teams' joint training to the table tennis championship in 1991 when the unified Korean team won the title. It should be noted that one of the staff of the South Korean table tennis team was the player who had won the women's doubles championship with a North Korean partner in 1991. Thus, the behaviour of the media which tried to build a linkage between the Olympics and the past table tennis championship in an effort to highlight a unified Korean identity was understandable. Overall, the inter-Korean relations at the Athens Olympic Games were some of the friendliest ever.

Co-operative relations continued after the Athens Olympic Games. North Korean support for the South Korean bid for the Winter Olympic Games exemplified this trend (Lee, 2006). The president of the North Korean NOC sent the IOC an official letter supporting Pyeongchang's campaign to host the Winter Olympic Games in 2014. This letter also claimed that if the Olympics were to be awarded to its southern neighbour, the North would seriously consider entering into a sporting union with the South for the Games. Both countries believed that if Pyeongchang hosted the sporting event, this would contribute to the process of gradual reunification. They also thought that the inter-Korean collaboration at the Olympic Games was a positive basis on which to build a peaceful and harmonious society through sport. In spite of their mutual support, the Pyeongchang Olympic bid was unsuccessful. Nevertheless, the political implications of the North's attitude towards the South should not be overlooked. Regardless of the result, the fact that the two Koreas had worked together for an international event can be perceived as progress in North and South Korean relations.

TWISTS IN BEIJING IN 2008

North and South Korean sporting relations became deadlocked in 2008 with a change in the political climate of the Korean peninsula. Before the second

Inter-Korean Summit in 2007, many Koreans expected more progress would be made, possibly resulting in a sporting union for the 2008 Beijing Olympics. In fact, at the second Summit the two sides agreed that they would at least form a united supporters' group, and that North Korea would temporarily open its railway to Beijing so that North and South Koreans could travel together. Moreover, the two NOCs held a number of bilateral talks to discuss creating a unified Olympic team. However, these co-operative efforts suddenly deteriorated into hostility with new political developments in the region.

In February 2008 a new conservative president was elected in South Korea and announced that he would take a tough approach towards North Korea, putting an end to the 'Sunshine Policy' of the previous two governments. In addition, the conservative regime confirmed a solid alliance between South Korea and the USA. The North Korean Government criticized the actions of the South Korean President, and attempted to end almost every economic and cultural tie established over the last decade. Inter-Korean relations appeared to reach their lowest point when a South Korean tourist was shot dead by a North Korean soldier at the special tourist zone in North Korea. Was this a sign of a new Cold War emerging on the Korean peninsula?

Inter-Korean sporting relations, especially the attempt to build a united Korean team for the Beijing Olympics, were directly affected by the worsening political circumstances (Lee and Maguire, 2009b). The agreement to create a united supporters' group that would travel together through North Korean territory by train was nullified as a result. The North and South Korean teams entered the stadium separately at the opening ceremony despite the IOC's attempts to make the joint march happen. North Korea even objected to entering the stadium immediately after the South Korean team. Instead, the North Korean team marched after the Austrian team. Unlike at the Athens Olympics when North and South Korean athletes interacted freely, the two teams in Beijing intentionally kept their distance. When high-level team officials occasionally encountered each other at the venue, they barely spoke. The Beijing Olympics marked one of the coldest Games in terms of inter-Korean relations.

CONCLUSION

The Olympic Games have been the most distinctive platform on which the two Koreas have displayed both unified and separate Korean identities depending on the broader political situation around the Korean peninsula. Given that the Korean peninsula is the last remnant of the Cold War, the Olympics have always had political implications for the two states. However, the competition for Olympic medals has no serious political meaning for the two Koreas in the post-Soviet era because they have barely used sport as a vehicle for touting the supremacy of their respective political ideologies. Instead, more political consideration is given to the way in which the Olympic teams demonstrate their unified or separate national identities at the venues, which depends on the political circumstances of the day. The two Koreas made some effort to show a unified Korean identity at the Sydney and Athens Olympic Games by marching together at the opening ceremonies. However, in Beijing the two sides displayed the coldest inter-Korean relations at any Olympics in the post-Soviet era. This suggests that North and South Korean relations are continually in flux and that wider political issues influence the behaviour of the two Olympic teams. Therefore, it seems that political factors constantly affect, if not determine, Koreans' Olympic experiences.

There is no doubt that sport has played a significant role in bridging a cultural gap between the two Koreas when political relations between them have signalled a mood of reconciliation. It was also in a sporting stadium that the most intense form of a unified Korean identity has been displayed. However, inter-Korean relations in sport have had a limited wider impact. In other words, there have been very few occasions when a sporting factor has directly facilitated the process of reconciliation in other areas. Moreover, such an amicable sentiment expressed through sport appears more or less volatile in nature. Subsequently, the feelings of an inter-Korean co-operation in sport tend to disappear easily when the context changes and political conflict develops. Therefore, the implication of sport for the promotion of mutually beneficial relations between the two Koreas should not be overestimated. On this basis, it can be argued that sport has mostly been a dependent variable on political factors at least in Korea.

REFERENCES

Chen, C. and J. Lee (2007) 'Making Sense of North Korea: "National Stalinism" in Comparative-Historical Perspective'. *Communist and Post Communist Studies* 40: 459–75.

Cummings, B. (2005) *Korea's Place in the Sun: A Modern History* (updated edn). New York: W.W. Norton & Co.

Larson, J. and H.S. Park (1993) *Global Television and the Politics of the Seoul Olympics*. Boulder: Westview Press.

Lee, H.S. (2006) 'Inter-Korean Sport Program Boosts Pyeongchang's Olympic bid'. *Korea Times*, 5 April.

Lee, J.W. and A. Bairner (2009) 'The Difficult Dialogue: Communism, Nationalism and Political Propaganda in North Korean Sport'. *Journal of Sport and Social Issues* 33(4), 390–410.

Lee, J.W. and J. Maguire (2009a) 'Global Festivals through a National Prism: the Global/ National Nexus in South Korean Media Coverage of the 2004 Athens Olympic Games'. *International Review for the Sociology of Sport Journal* 44(1): 5–24.

——(2009b) 'The Portrayals of North Korean Athletes at the 2004 Athens Olympic Games by the South Korean Newspapers'. *International Journal of Human Movement Science* 3(1): 5–29.

Merkel, U. (2008) 'The Politics of Sport Diplomacy and Reunification in Divided Korea: One Nation, Two Countries and Three Flags'. *International Review for the Sociology of Sport* 43(3): 289–311.

Rodong Shinmun (1985) 'Editorials'. Newspaper, 7 October.

Senn, A.E. (1999) *Power, Politics and Olympic Games: A History of Power Brokers, Events, and Controversies that Shaped the Games.* Champaign, IL: Human Kinetics.

Son, K. (2006) *South Korean Engagement Policies and North Korea: Identities, Norms and the Sunshine Policy.* London: Routledge.

FURTHER READING

Bridges, B. (2007) 'Reluctant Mediator: Hong Kong, the Two Koreas and the Tokyo Olympics'. *The International Journal of the History of Sport* 24(3): 375–91.

Ha, N.G. and J.A. Mangan (2003) 'Ideology, Politics, Power: Korean Sport – Transformation, 1945–92', in J. Mangan and H. Fan (eds), *Sport in Asian Society: Past and Present.* London: Frank Cass, 213–42.

Pound, R.W. (1994) *Five Rings Over Korea: The Secret Negotiations Behind the 1988 Olympic Games.* New York, NY: Little, Brown and Co.

Taiwanese identities and the 2008 Beijing Olympic Games

PING-CHAO LEE, ALAN BAIRNER AND TIEN-CHIN TAN

INTRODUCTION

In the context of the history of the Olympic Games, one very clear and persistent example of the interweaving of sport and politics is provided by the issue of the 'two Chinas'. According to an editorial in a Taiwanese English-language newspaper, 'Taiwan's isolation from the formal world community despite our substantive status as a democratic independent state due to opposition by the authoritarian People's Republic of China has been a long-standing and deeply felt injustice to our 23 million people' (*Taiwan Times*, 30 September 2009). In fact, one element of the world community from which Taiwan is not formally excluded is the Olympic Movement. Its involvement, however, has for many years been in circumstances chosen by others—the IOC and the People's Republic of China (PRC)—rather than by the people of Taiwan themselves. For that reason, the Olympics have long been recognized as political by the Taiwanese.

It was widely agreed that the opportunity for the PRC to host the Olympic Games in Beijing in 2008 would have crucial consequences for the political and economic development in the Asia-Pacific region and, indeed, the whole world (Yu and Mangan, 2008). It is also generally believed, despite the views of sceptics, that an Olympic host country can obtain significant political and economic benefits in terms national integration, facility construction, city maintenance and increased tourism as a result of holding the Olympics. For the PRC, the right to host the 2008 Olympic Games in Beijing was particularly significant for a number of reasons, not least as a symbol of the PRC's emergence as a modern, developed country. It also allowed for the boosting of Chinese nationalism and enhancing the state's legitimacy in the eyes of the Chinese public (Chan, 2002; Xu, 2006). However, how would the Beijing Games be viewed by the citizens of the Republic of China, more commonly known as Taiwan but, largely because of the influence of the PRC, obliged to operate under the name 'Chinese Taipei' in countless international organizations, not least the International Olympic Committee (IOC). This chapter examines Taiwanese responses to the attempt to boost Chinese nationalism by way of the Beijing Olympics. In addition, the chapter analyses the political confrontation and attempted compromises between the PRC and Taiwan in relation to the issue of the 'two Chinas' in relation to Olympic history in general, but with particular reference to the Beijing Games. Finally, the chapter examines the impact on Taiwanese identities of the PRC's staging of the 2008 Olympic Games within the context of recent developments in Taiwanese politics.

From the outset, in relation to the 2008 Beijing Olympic Games, Taiwan clearly identified itself as not being part of the host nation, the PRC. Indeed, this distancing was seen as having the potential to strengthen Taiwanese identity and demands for independence. In the event, however, the Games proved to be significant as a political tool for both Taiwan and the PRC. The latter demonstrated its goal of unification from the beginning of the Games preparation, not least in establishing the torch relay route, which is discussed below, but it was unable to achieve its ultimate goal in the face of opposition from the Taiwanese Government. The PRC's leaders may have hoped that Taiwanese national identity would come closer to its ideal of 'one China' nationalism. In this respect, however, they may well have been insufficiently aware of the complexity of Taiwanese national identity debates. Xu (2006) argued that the Beijing Olympics could play a constructive role in forming a Chinese identity across the Taiwan Straits. Given the findings of previous studies on the association between national identity and the Olympics or other mega sporting events, it is certainly worthwhile investigating the impact of the 2008 Olympic Games on changes, if any, in the national identity of the citizens of Taiwan. First, however, it is necessary to provide a brief account of the historical background to these contemporary debates.

'CHINESE IDENTITIES' AND THE 'TWO CHINAS' ISSUE IN THE OLYMPIC GAMES

According to Jarvie *et al.* (2008: 108), 'the two-Chinas issue may be seen as a conflict of national identities that was originally created by the actions of both the Chinese Communist Party (CCP) and KMT [Kuomintang] after 1949'. Subsequently, the political conflict between the PRC and the ROC brought the 'two Chinas' issue into the international diplomatic and sporting arenas. It specifically affected the Olympic Movement because both the PRC and Taiwan claimed to be the sole legitimate representative of China (Jarvie *et al.*, 2008). The KMT party was founded on the so-called 'one China' principle, which sustained the claim that there is only one China, that Taiwan is part of China, and that the ROC Government is the sole legitimate representative of the whole of China. Thus, with mainland Chinese politicians dominating the KMT, which in turn dominated Taiwanese politics, the main thrust of policy towards mainland China was the goal of reuniting Taiwan with China (in a post-communist system). In other words, returning to the mainland and rebuilding the country was the first priority of the KMT Government. For many islanders, however, the 'returning to the mainland' political objective meant nothing because they had not shared the same experience as the mainlanders during the civil war which led to the KMT decamping to Taiwan. The task of cultivating or imposing on the islanders a 'Chinese' consciousness was vital for the KMT Government and they attempted to pursue this in two main ways.

One method was using military force and enforcing the so-called 'White Terror' policies, martial law and 'Temporary Provisions Act' as the means by which to control the people and stabilize the KMT's political legitimacy. The second primary method was to utilize the education system to manipulate young people's beliefs and values, in order to forge a shared 'Chinese' identity. All political, economic and social policies were aligned to this goal, helping to explain why international sport (and in particular the relative positions of Taiwan and the PRC within the Olympic Movement) acquired huge symbolic importance as the issue of the 'two Chinas' began to manifest itself in the sports domain.

The 'Olympic formula' established by the IOC has allowed both the PRC and Taiwan to participate in the Olympic Games despite the fact that, unlike the PRC, Taiwan is not a member of the United Nations nor, indeed, as Taiwan or the ROC in the Olympic family. For Taiwan, there was no option but accept the resolution if it wished to stay in the Olympic Movement (Chan, 1985; 2002). The compromise was actually inspired by the doctrine of 'one country, two systems' proposed by Deng Xiaoping, thereby making it possible for communist China to change its attitude towards allowing Taiwan to take part in the Olympics on condition of using a name that posited the territory to be part of the PRC. Undoubtedly, the PRC was satisfied with the IOC's resolution as it solved the problem of legitimate and sovereign Chinese representation. In Taiwan, the 'compromise' was interpreted rather differently as is reflected in a statement made by the Minister of Education on Physical Education Day, 1982:

> Because of the NOC's title issue in the Olympic Movement, we were unable to participate in the Games for years. Now the issue has been sorted out, thus to participate in the Games and to win the highest prestige for 'our country' will be the most important task.
>
> (Chu, 1982: 5, translated by one of the authors)

Here the concept of our country [the ROC] is emphasized in response to China's political ambitions according to which the PRC hopes that the Olympic formula could act as a catalyst in the process of reunification of Taiwan with the mainland (Chan, 1985). None the less, this so-called 'Olympic formula' may well have actually weakened Chinese identity in Taiwan, not least because the PRC now officially represented China (Yu and Bairner, 2008). More precisely, Taiwan's authorities could show to the world that there existed a separate Chinese [political] entity in Taiwan in addition to the one in mainland China (Chan, 1985).

It is evident that between 1945 and 1991, Taiwan's government had portrayed Taiwan as ethnically Han and nationally Chinese, even claiming that it was the lawful government of mainland China (Brown, 2004). This position started to weaken in the early 1990s, increasingly replaced by one that does not challenge the legitimacy of the PRC's rule in mainland China, particularly since the first native-born Taiwanese President, Lee Teng-Hui, was elected. Thus the struggle to elucidate a specifically Taiwanese identity became far stronger and more explicit through the 1990s as President Lee provided the 'impetus for abandoning the KMT's core commitment to Chinese nationalism ... and facilitated ideological accommodation with opposition on the issue of democratic reform and national identity' (Chu and Lin, 2001: 121). Indeed, 'with the political and economic transformations of the 1980s, Taiwanese identity has changed dramatically, becoming increasingly inclusive, self-confident and nationalistic' (Brown, 2004: 12).

From this point on, the development of state policy in relation to Taiwan's political identity shifted from an agreed framing of the island as an integral part of mainland China to a tension between this view and an alternative policy of promoting Taiwan as an independent political entity. With the emergence of the Democratic Progressive Party (DPP) and its subsequent coming to power in 2000 and 2004, this policy tension became the major defining difference between the two major parties. For the DPP Government, sport (baseball in particular) was utilized as an important facet of claims to nationhood, and national pride on the part of the Taiwanese ('One China, One Taiwan'), intensified with the ongoing troubled relationship with mainland China.

THE CHALLENGE OF BEIJING

For the Chinese Government, hosting the 2008 Beijing Olympic Games was intended to give political and national meaning for the country, particularly in relation to the recognition of a successful communist leadership both in domestic and international arenas. In fact, whilst the award of the Games provided new impetus to China's modernization drive and international integration, it also forced the PRC to face some intractable contradictions in the modernization process. The vision of staging 'the best ever Olympic Games' would be subject to close examination centred on China's ability to deal with several major issues, amongst them the complex identity politics involved in containing the Taiwan independence movement and promoting greater national reconciliation across the Taiwan Strait (Xu, 2006). The Chinese Government selected the 2008 Beijing Olympics as a medium for promoting pride in Chinese national identity and values. As commentators argued, the notion of the 'one China principle' had become such an entrenched part of its sporting politics that the Chinese Government's attempt at 'territorial integrating action' would inevitably increase tensions between competing factions (Jarvie et al., 2008; Yu, 2008; Yu and Mangan, 2008). Two specific examples serve to illustrate the precise character of relations linked to 'Chinese identity' between the IOC, the PRC and Taiwan in this sport-politics nexus.

The torch relay has become an expression of universalism in the context of the host nation with an emphasis on using the route to encapsulate the peaceful dimensions of the world's multiple cultural, ethnic and political identities. With specific reference to Taiwan, the delicate political situation concerning the 2008 Olympic torch arose from the decades-old difference in standpoints between mainland China and Taiwan regarding state sovereignty claims. Because a torch relay route including Taiwan might imply that 'Taiwan is part of China, which Beijing so insists and Taipei so resists, it raises a serious doubt that this may happen at all' (Xu, 2006: 104). For a long time, China has regarded Taiwan as its provincial state and thus considered it as an appropriate 'local' stop in the mapping of that section of the Olympic torch's route, which would also include the PRC's Special Administrative Regions, Hong Kong and Macao. In response, Taiwan's government rejected Beijing's proposal to let the Olympic torch pass through Taiwan because this would amount to admitting that the nation is part of China. Consequently, Taiwan became 'the first IOC member to decline to be included in the route and accused China of undermining Taiwan's sovereignty by deliberately creating the impression that Taiwan was part of China' (Yu and Mangan, 2008: 838).

The second example relates to the title of Taiwan's Olympic team and the arrangements for the opening ceremony procession. Aware of the growing Chinese nationalistic atmosphere surrounding the 2008 Beijing Olympic Games which was expected to provide China with an opportunity to promote greater Chinese identity and integrate the pan-Chinese community, particularly in Taiwan, the latter's government sought to establish a 'firewall' to lessen the impact of the PRC's strategy. Taiwan (or Chinese Taipei as a member of the IOC) expected all of its rights to be confirmed and protected under the IOC Protocol Guide signed in 1981 between the Chinese Taipei Olympic Committee and the IOC. As a consequence, the Chinese Taipei Olympic Committee expressed Taiwan's anxiety to the IOC. Responding to Taiwan's concerns, in a letter to Kou-I Chen (Secretary-General of the Chinese Taipei Olympic Committee) and dated 15 July 2008, Pere Miro, the IOC's Director of NOC Relations said in relation to the Opening Ceremony Marching Order:

> Thank you for your letter dated 17th June 2008, concerning the Marching Order of Chinese Taipei for the Opening Ceremony of the Beijing Olympic Games 2008. The IOC is currently working closely with BOCOG to ensure that the correct Olympic Protocol is adhered to and that the Marching Order shall be in accordance with the IOC Directory, placing the Chinese Taipei in the alphabetic order of "T" under the name of Taipei. BOCOG is to send the provisional Marching Order to the IOC for the approval, at which time the IOC will review the proposed order to ensure that all the correct procedures have been observed and the arrangement of the Marching Order, even in the official language of that host country, does not depart from Olympic Protocol and past practices.
> (Chinese Taipei Olympic Committee, 2008a)

According to the response, the IOC clearly indicated that the Taiwanese team would be abbreviated as 'TPE' and thus listed under the 'T' section in the IOC's directory. This model has been used for years and applied to other sports events (Su, 2008). Nevertheless, Beijing decided to arrange the opening procession based on the number of strokes in the first character of each country or territory's name in simplified Chinese and thus the marching order of the Chinese Taipei delegation was arranged under C (for China) rather than T (Taipei). Thus Gilbert Felli, the Olympic Games Executive Director of the IOC, wrote to Chen on 28 July 2008 expressing the IOC's standpoint:

> According to the IOC Protocol Guide, in which it is stipulated that Marching Order should be based on the language of the host city, we have received from BOCOG the marching order for the NOCs in the athletes' parade for the Opening Ceremony of the Games of the XXIX Olympiad which you will find enclosed with the present letter. We would also like to confirm that for the next Games in Vancouver and London the alphabetic order will be in English and the Chinese Taipei Olympic Committee will fall under the letter T.
>
> (Chinese Taipei Olympic Committee, 2008b)

Clearly, the rhetoric from the Chinese Government 'attached great political importance to the Beijing Olympics in terms of constructing national identity and pursuing international primacy' (Xu, 2006: 104). While nationalism was expected to be inspired by the first Olympics in Chinese history, it is important to understand the degree to which the Taiwanese engaged in this process in terms of the construction of a 'Greater Chinese Community' which the PRC sought to achieve.

CHINESE OR TAIWANESE? IDENTITY ISSUES AND THE 2008 OLYMPIC GAMES

In the remainder of this chapter, based on a survey of Taiwanese collegiate athletes, we consider whether or not Chinese and Taiwanese national identities were changed by the 2008 Beijing Olympic Games. During the research process, we first asked our research participants before the 2008 Beijing Olympic Games to indicate how their national identity might be affected. We then asked interviewees after the Beijing Games if the Games had indeed affected their sense of national identity.

This study began in January 2007 and involved questionnaire surveys and interviews conducted before and after the Games. The initial research method employed a questionnaire survey approach to collect information which helped identify respondents' positions on Taiwanese identity in relation to the circumstances of the 2008 Beijing Olympic Games. Research participants were identified using the 2007 National Intercollegiate Athletic Games in Taiwan as the sampling frame. These Games included competitions across 11 sports and involved 8,199 student athletes from 167 universities in Taiwan. The collegiate athletes were selected for our enquiries because of their status as both athletes and citizens. In addition, their high level of education, together with their knowledge of both sports affairs and the particular relationship between China and Taiwan were also important concerns. They represented a group of Taiwanese citizens who would be most likely to respond favourably to research inquiries on this issue. Data were collected from 4 May to the end of July 2007 because most of the subjects did not finish/return the questionnaires themselves and questionnaires were returned by their coaches after the Games. A total of 2,176 surveys, representing 23.5% of the total number of respondents, were conducted and 1,929 surveys were utilized in the subsequent analysis.

The study additionally drew on interviews conducted after the Beijing Games when interviewees' accounts of their thinking about the Games provided the basis for a deeper analysis of the relationship between Beijing 2008 and national identity issues. In total, 50 interviews with collegiate athletes who participated in the 2007 National Intercollegiate Athletic Games were undertaken between 1 September and 3 October 2008 by the first author. Interviews were subsequently transcribed and translated by two of the authors.

At the end of 2006, a survey instigated by one of the major local media, *United Daily News* (*Lianhebao*, 聯合報), which leans politically towards unification with China, showed that the 2008 Beijing Olympics seemed unlikely to stimulate Taiwanese 'resonance of Chinese nationalism' as only 14% of Taiwanese regard the event as a matter of prestige and pride (in the Chinese state). In the present study, Taiwanese attitudes towards Beijing's hosting the 2008 Olympic Games are reported in Table 10.1, which reveals that nearly 100% of the respondents knew that Beijing was to be the host city. When the respondents were asked how they 'felt' about Beijing hosting the Games, 29.3% of them were pleased and 51.2% had no opinion. More than half of the respondents reported no opinion, thus making it impossible to know what their real views were.

The Beijing Olympics were closely linked to identity politics over the status of Taiwan in diverse ways. These figures reflect the fact that the response of the Taiwanese to the news of the successful Beijing Olympic bid was mixed. The ambivalence persisted during the Games as one interviewee highlighted:

Table 10.1 Taiwanese attitudes towards Beijing hosting the 2008 Olympic Games

	Knowing that Beijing is going to host the 2008 Olympic Games	*Proud of Beijing hosting the 2008 Olympic Games*
Yes	1,917 (99.38%)	803 (29.30%)
No	12 (0.62%)	209 (19.50%)
No opinion/ other	0 (0.00%)	917 (51.20%)

China did so well in the Olympic Games. I was impressed by its splendid opening ceremony, which was able to demonstrate Chinese culture in a modern way. This pride is definitely welcomed by the Chinese government and people; however, it is not meaningful for me. I only counted on the performance of Taiwan's athletes ... unfortunately their performance [four Bronze medals] did not satisfy Taiwan's people. Meanwhile, people were also angry and disappointed that our baseball team lost to China in a dramatic ending.

(Interviewee K, 18 September 2008, translated by one of the authors)

The circumstances surrounding the collision between political estrangement and cultural acceptance illustrate the complexity of Taiwan's citizens' identities. The use of 'Chinese Taipei', which has represented Taiwan in many international instances in recent decades, also prompts oscillation of the citizens' identities spanning reluctant acceptance, acceptance and refusal. In order to assess Taiwanese identity perceptions regarding China hosting the 2008 Olympic Games, the survey asked respondents a number of questions. The opening question was intentionally vague and sought to discover the understanding of Taiwan's official title in the Olympic Movement, with respondents subsequently being asked to what extent they were willing to accept such a title. The following questions sought attitudes to three possible scenarios.

Table 10.2 shows that more than two-thirds of the respondents, or 1,285 (66.61%), understood that the name of 'Chinese Taipei' is currently employed as the official title for Taiwan when participating in the Olympic Games. Interestingly, when the participants were further asked about their preferences for Taiwan's name in the context of the Olympic Games, a range of opinions emerged. According to the data, Taiwan's current title, 'Chinese Taipei' (27.37%), is not the respondents' preference, as the name of Taiwan (37.07%) occupied first position and Republic of China (at 26.18%) was only marginally behind 'Chinese Taipei'. The title of 'China Taipei' (1.24%), which the PRC Government preferred to use at international and domestic levels, was the least popular among the Taiwanese.

Table 10.2 Taiwanese attitudes towards Taiwan's title in the Olympic Games

	What is Taiwan's current official title when participating in the Olympic Games?	What would you prefer as Taiwan's official title when participating in the Olympic Games?
Republic of China	206 (10.68%)	505 (26.18%)
Taiwan	287 (14.88%)	715 (37.07%)
China Taipei	89 (4.61%)	24 (1.24%)
Chinese Taipei	1,285 (66.61%)	528 (27.37%)
No opinion/ other	62 (3.22%)	157 (8.14%)

Table 10.3 lists respondents' attitudes towards Taiwan's name issued in relation to athletes participating in the 2008 Olympic Games, for which the PRC had its own clear preference. Overall, 'China Taipei' was viewed by respondents (35.77%) as the least preferred title with which to introduce athletes, followed by 'Chinese Taipei' and Taiwan, chosen by 30.27% and 21.88% of the participants, respectively. In addition, only around one in 10 (10.63%) of the respondents agreed with the title Republic of China, which was perceived as problematic because of the 'Two Chinas' obstacle. While the analysis gives an indication of Taiwanese 'consideration' towards China's possible actions at the Beijing Olympic Games, concerns in this respect were also raised by the respondents' most and least preferred choices and their understanding of 'adequate solutions' and 'title conflict' between the two sides. Indeed, the research found that there was a discrepancy between respondents' preferences and their 'notional setting' with reference to China's possible actions. Some 38.67% of respondents expressed the view that 'Chinese Taipei' could be utilized by the host to indicate from where the Taiwanese athletes came, while almost one-third (32.14%) of respondents preferred the name Taiwan. The designation deemed unacceptable for introducing Taiwan's athletes, by more than two in three respondents (66.98%), was 'China Taipei', which respondents perceived as inappropriate, as Beijing had deliberately used this name as a means of demonstrating its claim of sovereignty over Taiwan.

While the Olympic formula seems to have worked so far whereby both China and Taiwan simultaneously participate in the Olympic Games as well as other non-sport organizations, the 2008 Beijing Games posed new questions. These findings, in conjunction with results shown in the previous tables, reveal that the Taiwanese people were divided in terms of their support for Taiwan and Chinese Taipei as a title for participation in the international sporting arenas, while the usage of China Taipei was largely unacceptable. In the opinion of most respondents, China's claim to Taiwan, considering

Table 10.3 Taiwan's athletes' participation in the 2008 Olympic Games

	When Taiwan's athletes arrive in the 2008 Beijing Olympic Games stadium, where should the hosts introduce them as being from?	*Title for Taiwan's athletes' place of origin that you prefer for the 2008 Beijing Olympic Games*	*Title for Taiwan's athletes' place of origin that you refuse to accept for the 2008 Beijing Olympic Games*
Republic of China	205 (10.63%)	313 (16.23%)	95 (4.93%)
Taiwan	422 (21.88%)	620 (32.14%)	67 (3.47%)
China Taipei	690 (35.77%)	30 (1.56%)	1,292 (66.98%)
Chinese Taipei	584 (30.27%)	746 (38.67%)	133 (6.89%)
No opinions / other	28 (1.45%)	220 (11.40%)	342 (17.73%)

Taiwan to be a part of China, was inconsistent with the political reality underpinning Taiwanese consciousness—namely, that of *de facto* sovereignty over a number of decades (Li, 2006). As Brown (2004: 28) notes, 'Taiwan has been governed separately from China since 1949—but the rhetoric of different culture and ancestry has not been repudiated'. After nearly 60 years of separation, many differences have emerged in terms of political, social and economic development, which are the key elements of division between the two sides. Nevertheless, the overwhelming majority of Taiwan's people still share Chinese ancestry and culture. Curiously, the new generation of Taiwan's citizens—those born in the 1980s—seem to be culturally closer to China than ever before, but even more estranged from it politically.

When the respondents were asked how they would cheer for the Taiwanese athletes at the Beijing Games (Table 10.4), the slogan 'Go Go Republic of China' (Go Go Chinese Team, *Zhonghua Dui Jia You*) was chosen by 47.07%, whereas the proportion that opted for 'Go Go Taiwan' (*Taiwan Dui Jia You*) constituted 41.47% of the total. This result reveals that the continuing co-existence of Chinese nationalism (promoted by the KMT) and the emerging Taiwanese nationalism (promoted by the DPP) continues to affect how the Taiwanese shape their national identities. One major concern here is that Taiwan's residents identify 'Chinese Team' (*Zhonghua Dui*) as the national team of the ROC (Taiwan), and used 'China Team' (*Zhongguo Dui*) to refer to the representatives of the PRC. In the Mandarin language, the two words, 'Chinese' and 'China' could be interpreted as referring to two different political entities, although they imply a shared Chinese ancestry. Consequently, 'drawing a sharp distinction between culture and politics, Taiwan continues to affirm its 'Chinese' linguistic and cultural heritage while claiming full political autonomy' (Dittmer, 2004: 480). Indeed, it is the intertwining of the cultural and political components that constitutes the very essence of public discourse on national identity in Taiwan.

HAS THE 2008 BEIJING OLYMPIC GAMES AFFECTED TAIWANESE IDENTITY?

Houlihan (2004: 213) has pointed out that 'sport can be an important ingredient in the politics of identity'. In the case of Taiwan, sport is also bound up

Table 10.4 Taiwanese attitudes towards their national team in the 2008 Olympic Games

	When Taiwanese athletes are competing in the 2008 Beijing Olympic Games, you will cheer for them with 'Go Go ...'?
Republic of China	908 (47.07%)
Taiwan	800 (41.47%)
Taipei	76 (3.94%)
No opinion/ other	145 (7.52%)

Note: Corresponding frequencies are followed by percentages in parentheses

in what might be termed 'the politics of contested identities'. When the respondents were asked, regardless of Beijing hosting the 2008 Olympic Games, if they would identify themselves as Taiwanese, both Taiwanese and Chinese, or Chinese (Table 10.5), the data illustrate that 62.41% selected Taiwanese as their preference; 27.06% preferred both Taiwanese and Chinese, and only 8.09% selected Chinese as their favoured mode of self-identification.

These figures are consistent with earlier results in relation to preferred national identities which were predominantly Taiwan-centred with China being seen by most respondents as a separate polity. This reveals the extent to which the situation has changed since the introduction of democratic principles and practices to Taiwan. In the face of China's intransigence (e.g. the one China principle), it is generally accepted that fewer and fewer Taiwanese are likely to identify themselves as Chinese (Yu and Mangan, 2008). Indeed, according to earlier surveys carried out by National Chengchi University in 2006, the number of citizens who consider themselves Taiwanese has nearly doubled in recent years to 44%, while the figure for those who classify themselves as Chinese has dropped from 16% to 6% over the past decade (Osnos, 2007). However, despite an apparent trend towards embracing a Taiwanese national identity, interestingly around 57.39% of respondents agreed that Taiwan's culture is part of Chinese culture, thereby implying that Taiwan and China's cultures are inseparable.

The push for a distinctive Taiwanese identity has many sources. It is in part a reaction to posturing from the PRC and, consequently, an attempt to justify an existence politically different from its neighbour. It is also a reaction to many decades of heavy-handed cultural control by the KMT, one goal of which was precisely to weaken any sense of Taiwanese consciousness in favour of a more generic Chinese identity. For example, the KMT used various tactics in an effort to create a sense of Chinese identity, such as commemorating national holidays, installing standardized textbooks at school, and promoting cultural events, all of which were intended to create an illusion of cohesion (Yu and Bairner, 2008). Nevertheless, 'identity change did not occur without a change in the political regime' (Brown, 2004: 232). In Hong

Table 10.5 Taiwanese national identity towards the 2008 Olympic Games

	Without considering the fact that Beijing is hosting the 2008 Olympic Games, will you identify yourself as:	*In the light of Beijing hosting the 2008 Olympic Games successfully, will you identify yourself as:*
Chinese	156 (8.09%)	239 (12.39%)
Taiwanese	1,204 (62.41%)	1,042 (54.02%)
Both Chinese and Taiwanese	522 (27.06%)	576 (29.86%)
Neither	47 (2.44%)	72 (3.73%)

Note: Corresponding frequencies are followed by percentages in parentheses

Kong's case, recent developments have converged in the direction of re-Sinicization and the outstanding sporting achievements of elite athletes from China and Hong Kong have been instrumental in helping to construct cohesion based on Chinese identity between the two (Lau *et al.*, 2008). In Taiwan, on the other hand, during the 2001–08 period, the DPP Government sought to counterbalance decades of Sinicization and successfully highlighted national identity issues, promote Taiwanese identity and distance the island from its increasingly powerful neighbour. Recurring political participation under a democratic regime has helped develop a sense of collective consciousness among people, transforming the term 'Taiwan' from merely a geographic descriptor to a political community and the term 'Taiwanese' from an ethnic description of native Taiwanese to a civic term referring to all of the citizens of Taiwan (Chu and Lin, 2001). Thus, 'culture and ancestry are not what ultimately unite an ethnic group or a nation. Rather, identity is formed and solidified on the basis of common social experience, including economic and political experience' (Brown, 2004: 2).

Commentators have agreed that the 2008 Beijing Olympic Games was far more than a sporting event; rather it was a specific socio-political experience in Chinese history (Mangan, 2008; Xu, 2006). In the lead up to the Games, according to Yu and Mangan (2008: 845), 'many Taiwanese also consider that success in presenting and performing the games has become vital to China's global self-esteem'. Thus, it was not hard to imagine that the PRC would take advantage of the Olympics to promote Chinese nationalism. Some commentators further claimed that 'the danger of the Beijing Olympic Games is that it allows the popularity of sport to whip up spontaneous nationalism in China and this patriotism becomes the logic for state nationalism and a move by the PRC to advance on Taiwan' (Jarvie *et al.*, 2008: 113).

To evaluate the Beijing Games' political influence, respondents were asked to identify themselves in the event of Beijing hosting the Olympics successfully (Table 10.5). Respondents who claimed they were Chinese constituted more than 12%, with an increase of 83 people (4.30%), from 156 (8.09%) to 239 (12.39%). The option of Taiwanese still came first but its ratio declined from 62.41% to 54.02%. In addition, the percentage of the group identifying as both Chinese and Taiwanese showed a slight increase, from 522 (27.06%) to 576 (29.86%). This confirms Brown's assertion that 'identities are both fluid and changeable … allowing individuals and even entire groups to fall first on one side and then on the other side' (Brown, 2004: 16–17).

According to the results above, a successful hosting of the 2008 Olympic Games was capable of affecting Taiwanese self-identification, albeit not to a dramatic extent. In light of respondents' potential for changing their national identification, the Olympics certainly appears to have given China an ideal opportunity with which to undermine confidence in and identification with a distinctive Taiwanese identity. Nevertheless, it is unlikely that Beijing hosting the Olympic Games will be the catalyst for any instant or major transformation of Taiwanese political attitudes.

In Taiwan's case, the 'state' has been used in recent years to foster the growth of Taiwanese nationalism and to consolidate the 're-imagined community' both at home and in the international community (Chu and Lin, 2001), thereby successfully generating a new and unique 'Taiwanese' (as opposed to 'Chinese') identity. Thus, it is not surprising that, despite Beijing having successfully hosted the 2008 Olympic Games, only a minority of respondents considered replacing their Taiwanese national identity with a Chinese identity. Meanwhile, the PRC Government expects 'citizens' (including Taiwanese, Tibetans, Uygur, etc.) to subordinate their individual national identities to China's national ones. Reflecting on this point, respondents who were interviewed were keen to distinguish their own response to the Olympic Games from the expectations or aspirations of the PRC:

> The Chinese team won many gold medals and topped the ranking list. Good for the Chinese government and people ... however, China's success definitely can't change my personal political attitude as this pride belongs to China but not Taiwan ... the fact is that Taiwan has its own national team and we Taiwanese should support it.
> (Interviewee D, 8 September 2008, translated by one of the authors)

> I must admit that Beijing hosted the Olympic Games successfully ... but this will not influence my Taiwanese identity one little bit. It would be hard for me to think of becoming Chinese to celebrate Beijing's effort and success, even though we share the same Chinese cultural ancestry. In my mind, China is a big nation, which is our neighbour. My country is Taiwan and my compatriots are Taiwanese.
> (Interviewee Y, 24 September 2008, translated by one of the authors)

In 2001, when Beijing comfortably beat Paris, Toronto, Osaka and Istanbul to win the right to host the 2008 Olympic Games, there was concern that the PRC would use the Olympics to promote a stronger sense of Chinese identity, thus allowing the Taiwanese to feel proud as Chinese citizens. Taiwan's former Vice-President Lu responded to this concern and pointed out that 'Beijing's hosting the 2008 Olympic Games will greatly challenge the ethnic and national identification of the Taiwanese people' (cited in Xu, 2006: 101). Moreover, another commentator claimed that:

> The Beijing Games are significant in several ways ... if Taiwan can be persuaded to become actively involved in organizing some parts of the Games, then China can claim credit for making progress in reunification, thereby boosting Chinese nationalism and enhancing its legitimacy in the eyes of the Chinese public.
> (Chan, 2002: 142)

The suggestion of co-hosting the Olympic Games by the PRC and Taiwan and the torch relay route were both affected by the 'One China' principle.

Meanwhile, the 'psychological impact' of the successful Beijing Games, the rapid rise in Chinese nationalism, has stimulated a resurgence of Chinese identity amongst mainland and overseas Chinese. Nevertheless, the majority of Taiwanese people were largely unaffected. Despite the success of the 2008 Beijing Olympic Games, the failure to reconcile political differences has been contrary to what Beijing intended—for example, in terms of promoting a common Chinese national identity across the Taiwan Strait.

CONCLUSION

This study has attempted not only to investigate the connection between the Olympic Movement and politics in Taiwan, but also to provide some theoretical and empirical findings that can inform and shed further light on the broader debate on the relationship between sport, politics and national identities. Retrospectively, from being internationally recognized as the official government of the whole China, temporarily in exile on the island of Taiwan, to being expelled from the United Nations and having diplomatic recognition withdrawn by all but a handful of smaller countries, the ROC/KMT Government had many challenges to face. From the problems of effectively being the colonizer of Taiwan and using the island as a garrison in which to regroup for an attack upon the communists in China, to bringing together, into one cohesive unit, a disparate group of peoples and forming them into one 'nation' with a new and shared identity, and to dealing with the problems of international diplomatic isolation, the Government recognized from the outset that sport could play an important role. Within the context of the Olympic Movement, the KMT administration utilized sport as a vital element in its strategy of fostering Chinese [Republic of China/Taiwan] national identity for political purposes. Although the Olympic Movement is a global phenomenon, it has, nevertheless, operated within a specific regional environment in the Chinese context and became a battlefield for the enactment of contestation between two different political entities, each claiming Chinese sovereignty and identity.

The Taiwanese political system has modernized with the shift from authoritarian rule to that of multi-party democracy. However, the challenge of maintaining power by that critical mixture of consent and coercion, thereby maintaining the 'agreement' of the governed people, is perhaps greater now than ever before precisely because of the democratic nature of politics. The struggle to develop a specifically Taiwanese identity has become more explicit and far more intense through the 1990s. Coupled with the promotion of some key political goals such as the establishment of a sovereign and an independent state by the DPP Government, there is the recognition of how a growing sense of Taiwan's sovereignty and national identity is changing and shaping public attitudes. The 'consciousness' amongst the general public of a 'Taiwanese nation', supported by the DPP, began to grow and becoming stronger than the 'consciousness' of the 'ROC nation' promoted by the KMT.

The ideology of what constituted 'Taiwan nationalism' began to significantly challenge the KMT's hegemony which, in turn, has had a profound impact on sport development. As more and more Taiwanese feel little or no connection to mainland China, it has become less and less likely that people would be profoundly influenced by the 2008 Beijing Olympic Games. Nevertheless, although in much of Taiwanese society people distance themselves from Chinese national identity, they acknowledge and honour their Chinese heritage. To that extent, there is evidence that the 2008 Beijing Olympic Games strengthened Chinese cultural identity among Taiwanese citizens.

According to Dittmer (2004: 483), 'research continues to be driven by events, the national identity issue in Taiwan has been driven by very contemporary political realities, which will no doubt continue to inform the research agenda'. In this respect, the process of making sense of the relationship between sport and politics in general, and between the Olympic Movement and national identity in particular, is a project which is far from completion— in Taiwan as elsewhere.

REFERENCES

Brown, M.J. (2004) *Is Taiwan Chinese? The Impact of Culture, Power, and Migration on Changing Identities.* Berkeley, CA: University of California Press.

Chan, G. (2002) 'From the "Olympic Formula" to the Beijing Games: Towards Greater Integration across the Taiwan Strait'. *Cambridge Review of International Affairs* 15(1): 141–48.

——(1985) 'The two Chinas problem and the Olympic formula'. *Pacific Affairs* 58(3): 473–90.

Chinese Taipei Olympic Committee (2008a) *Chinese Taipei opening ceremony marching order (I).* Taipei: Taipei Olympic Committee.

——(2008b) *Chinese Taipei opening ceremony marching order (II).* Taipei: Taipei Olympic Committee.

Chu, H.S. (1982) 'The speech of Minister of Education, Ministry on Physical Education Day'. *Physical Education Quarterly* 88: 1–3.

Chu, Y.H. and J.W. Lin (2001) 'Political development in twentieth century Taiwan: State-building, regime transformation and the construction of national identity', in R.L. Edmonds and S.M. Goldstein (eds), *Taiwan in the Twentieth Century: A Retrospective View.* Cambridge: Cambridge University Press, 102–29.

Dittmer, L. (2004) 'Taiwan and the issue of national identity'. *Asian Survey* 44(4): 475–83.

Houlihan, B. (2004) 'Practical policy issues', in D. MacIver (ed.), *Political Issues in the World Today.* Manchester: Manchester University Press, 212–25.

Jarvie, G., D.J. Hwang and M. Brennan (2008) *Sport, Revolution and the Beijing Olympics.* Oxford: Berg.

Lau, W.C., H.S. Lam and W.C. Leung (2008) *National Identity and the Beijing Olympics: School Children's Responses in Mainland China, Taiwan and Hong Kong.* Hong Kong: David C. Lam Institute for East-West Studies.

Li, C.P. (2006) 'Contending identities in Taiwan: Implications for cross-strait relations'. *Asian Survey* 46(4): 597–614.

Mangan, J.A. (2008) 'Preface: Geopolitical games – Beijing 2008'. *International Journal of the History of Sport* 25(7): 751–57.

Osnos, E. (2007) *Drops China from state institutions*. www.chicagotribune.com/news/nationworld/chi-0702200235feb20,1,4809343.story?page=1&cset=true&ctrack=1&coll=chi-newsnationworld-hed (accessed 14 February 2008).

Su, Y.Y. (2008) *Chi Cheng urges boycott of Olympic Games' opening*. www.taipeitimes.com/News/taiwan/archives/2008/08/02/2003419215 (accessed 14 February 2008).

Taiwan Times 2009, 'Editorial', Taipei, 30 September: 6.

Xu, X. (2006) 'Modernizing China in the Olympic spotlight: China's national identity and the 2008 Beijing Olympiad'. *Sociological Review* 54 (Supplement 2): 90–107.

Yu, J. (2008) 'China's foreign policy in sport: The primacy of national security and territorial integrity concerning the Taiwan question'. *The China Quarterly* 194: 294–308.

Yu, J. and A. Bairner (2008) 'Proud to be Chinese: Little league baseball and national identities in Taiwan during the 1970s'. *Identities: Global Studies in Culture and Power* 15(2): 216–39.

Yu, J. and J.A. Mangan (2008) 'Dancing around the elephant: The Beijing Olympics – Taiwanese reflections and reactions'. *International Journal of the History of Sport* 25(7): 826–50.

FURTHER READING

Corcuff, S. (ed.) (2002) *Memories of the Future: National Identity Issues and the Search for a New Taiwan*. Armonk, NY: M.E. Sharpe.

Lin, C.Y. (2003) *Taiwan Sport: The Interrelationship Between Sport and Politics Through Three Successive Political Regimes Using Baseball as an Example*. Unpublished PhD thesis, Eastbourne: University of Brighton.

Roy, D. (2003) *Taiwan: A Political History*. Ithaca, NY: Cornell University Press.

Rubinstein, M.A. (ed.) (2007) *Taiwan: A New History*. Armonk, NY: M.E. Sharpe.

Yu, J. (2007) *Playing in Isolation: A History of Baseball in Taiwan*. Lincoln, NE: University of Nebraska Press.

A–Z Glossary

Alan Bairner and Gyozo Molnar

A

Aboriginal people

The relationship between indigenous peoples and modern sport is problematic. On the one hand, involvement in modern sport is taken by some as evidence of assimilation. Alternatively, sport can be viewed as offering opportunities to native people denied to them in other spheres of life. Compounding this dilemma is the fact that racial discrimination represents a significant barrier to aboriginal sporting success in most parts of the world. The success of Cathy **Freeman** at the **Sydney Games** was in many ways the exception that proves the rule. **Protests** against the Games were arguably more significant in terms of drawing attention to the plight of aboriginal people in Australia as, indeed, are the 'Black Olympics' held each year since 1962—a sporting event organized by and for aborigines. Similarly, whilst only one competitor of aboriginal descent, Monica Pinette, represented Canada at the **Athens Games** in 2004, separate sporting events in North America ensure the maintenance of a distinct native athletic identity. Arguably, the two most famous aboriginal Olympians both represented the USA. Jim Thorpe (Wa-Tho-Huk) won gold medals in the pentathlon and the decathlon at the **Stockholm Games** in 1912. He was subsequently found guilty of professionalism and stripped of his medals. Not until 1982 were the medals posthumously returned to Thorpe's family and his name restored to the record books. In 1964, at the **Tokyo Games**, the 10,000 metres was won by Billy Mills, a Sioux who was raised on the Pine Ridge Indian Reservation in South Dakota. A film of his life—*Running Brave*—was made in 1984. Mills is co-founder and spokesperson of Running Strong for American Indian Youth, an organization that supports projects that benefit native people, especially the young.

At the **Beijing Games** in 2008, not only did Chinese Taipei (Taiwan) draw heavily on the island's indigenous population for its baseball team as has become customary, but Taiwanese aboriginals were also invited by the Chinese organizers and permitted by the Chinese Taipei Olympic Committee to take part in the opening ceremony. The extent of aboriginal involvement, whether athletic or ceremonial, at the 2010 **Winter Games** in Vancouver, Canada, remains to be seen.

Achievement principle

The achievement principle is the dark side of *Citius–Altius–Fortius* (Faster–Higher–Stronger). To conform to this idea and, indeed, to become faster and stronger and to achieve higher athletic goals require excessive training and the manipulation of human bodies, in the process of which athletes increasingly push the limits of human biological potential. By doing so, they risk their health, sacrifice their social life and may become what Alan Ingham called, 'specialized freaks'. By wanting to conform to the idea of *Citius–Altius–Fortius*, athletes allow their lives to be organized by a cadre of experts, amongst them coaches, physiologists, psychologists and dietetics, both on and off the sporting field. As a pragmatic response to the achievement principle, athletes are constantly under pressure to improve their performance, reach their very best and push their bodies beyond normal biological barriers. This demand, however, does not solely come from the ideology of modern sports, in general, and modern Olympics, in particular. There is another significant source of pressure. This is the athletic feeder system, through which new athletic talent is identified and developed with the ultimate goal of eventually surpassing current sporting achievements, thereby making the sporting personnel of today disposable heroes and heroines.

Amateur code

The argument that athletes in ancient Greece were all amateurs is a myth generated in the 19th century. In fact, no ancient Olympic athlete would have qualified for the modern Games strictly based on the principles of amateurism. Moreover, some scholars state that the ancient Greeks did not even have a word for amateurism as understood in the modern era. Others argue that *idiōtēs* is the ancient Greek equivalent of *amateur*, although the word actually means 'untrained', 'layperson' or simply 'non-competitor'.

The concept of amateurism, as we know it, has its roots in the 19th century and was widely promoted by the founders of modern sports and specifically the Olympics, thereby ensuring that the class-based exclusivity of certain sports and events was perpetuated. The amateur was usually a member of the aristocracy who participated in sport for its own sake, whereas professionals were normally working class and were involved in sports primarily for financial gain. In contrast to professionals, amateurs were expected to be solely concerned with their own pleasure, which they obtained by taking part in sporting activities. They even enjoyed the luxury of sacrificing results (e.g. breaking a record) in favour of self-gratifying, beautifully executed techniques. This reflects the original meaning of the French word, *amateur*, meaning 'a lover of', and refers to a voluntary and selfish motivation to partake as a result of a personal passion for a particular sporting activity.

Amsterdam Games (1928)

Amsterdam was first suggested as host city for the Olympics as far back as 1912, which was problematic given that there was no Netherlands Olympic Committee in existence at that time. Because of the outbreak of the First World War, the Olympic Games were suspended after the 1912 **Stockholm Games** until 1920. When the **International Olympic Committee (IOC)** finally met in Lausanne in 1919 to discuss where the Games of the VII Olympiad should take place, the Netherlands Olympic Committee abandoned its proposal to host the Games in favour of war-ravaged Antwerp, which was keen to organize the first post-war Olympics. Baron van Tuyll, the President of the Netherlands Olympic Committee, nevertheless proposed Amsterdam as the location for the Games in 1924, although several other cities were also proposed for that year. Because of his personal desire to commemorate the 30th anniversary of the Olympics, Pierre de **Coubertin** sought to ensure that the right to host the VIII Olympiad went to Paris. Therefore, in 1921, at the next IOC meeting, it was suggested that Paris and Amsterdam should be selected as the venues of the VIII and IX Olympic Games, respectively. Despite the IOC's decision, the American Olympic Committee had not given up on its ambition to host the IX Olympiad. Its delegation raised the possibility of Los Angeles hosting the 1928 Games at two consecutive IOC meetings and it was only courtesy of tactful political manoeuvring on the part of the Netherlands Olympic Committee that Amsterdam managed to retain its host city status.

The Amsterdam Games, comprising 15 sports and 109 events, were the Olympics of many 'firsts'. For instance, they were the first to be called the 'Summer Olympic Games', the Olympic Flame was lit during the Games, the parade of nations started with Greece and ended with the host country, Germany returned for the first time since 1912, athletics events were held on a 400-metre track that would become the standard length for athletics tracks, and these Games were the first to feature a standard schedule of 16 days (previously, the Olympic timetable had been stretched out over several months). These initiatives became integral to and remain part of the Olympic Games. Moreover, these were the Games at which the Coca-Cola company first appeared as sponsor. The Amsterdam Games are also famous for the women's 800 metres debacle. This was the first occasion when the event was included in the Olympic programme and most of the competitors were unprepared for the distance. Consequently, of the 11 competitors, only six completed the race and most of them collapsed after crossing the finishing line. Giving weight to the dominant male ideology of the time, female athletes' under-preparedness was interpreted as signifying that women were obviously biologically unfit for taking part in such physically gruelling events.

Through athletic success at the Games, some (mostly male) athletes achieved celebrity status. These included Johnny Weissmuller, who won two gold medals in swimming and later appeared in several Tarzan movies, and Paavo **Nurmi** who won his ninth and final Olympic gold medal in the 10,000 metres race, thereby reinforcing his social standing in Finland.

Another noteworthy feature of the Amsterdam Games was the distribution of medals. Athletes were summoned by loudspeakers and all first prize winners mounted the platform in front of the Royal Box, passing the Queen, together with other dignitaries, from whose hands they received their gold medals. The second prizes were distributed by the Prince Consort of the Netherlands and the third prizes by Count de **Baillet-Latour**, the IOC President.

Ancient Olympics

The earliest records of the ancient Olympic Games (*Olimpikos Agones*) date from 776 BC, although the precise origin of the Games is unknown. The Ancient Games—the most prestigious cultural event of ancient Greece—were held at Olympia, a religious sanctuary with numerous associations with Zeus, the ruler of the gods and to Mount Olympus, where a large temple, containing a 13-metre high gold and ivory statue, was erected in his honour. Due to the religious connotations of the Games, it was important for athletes to excel. Before the actual events (a five-day programme), athletes, their male relatives, trainers and judges had to take an oath before the statue of Zeus, promising to compete in an honourable way and to abide by the rules. Cheating was strongly discouraged and anyone found guilty would be heavily fined.

To compete in the ancient Olympic Games, athletes had to be male, of Greek origin and free. Women, slaves, criminals and foreigners were not allowed to compete. However, women did take part in running events in their own games (*Heraia*), which honoured Hera (the wife of Zeus). Free men, slaves and unmarried women were all allowed to watch the ancient Olympic Games, but married women were forbidden from attending.

In 146 BC, when Greece was conquered and became part of the Roman Empire, the ancient Games, after a brief revival, began to steadily decline. The rise of a monotheist religion, Christianity, also conflicted with the ancient Greek polytheistic world view. Finally, in AD 393, the Roman Emperor Theodosius I, who had made Christianity the official religion throughout his Empire, abolished the ancient Games.

Antwerp Games (1920)

Initially, the rights to host the VII Olympiad were awarded to Budapest, Hungary. However, because of the outcome of the First World War and the fact that the Austro-Hungarian monarchy had been a German ally during the conflict, the Games were transferred to Antwerp, Belgium, in 1919 at the first post-war IOC meeting. (Germany and its allies were not allowed to take part in the Antwerp Games.) Although other countries had bid to host the Games of the VII Olympiad, some decided to set aside their ambitions in favour of the war-ravaged, severely bombarded Antwerp, which was keen to host the first post-war Games. As the Games were transferred to Antwerp in 1919, the city, still struggling with post-war hardship, had only a short time to make the necessary arrangements required. Despite the hasty preparation, austere socio-economic conditions and post-war political tensions, 29 countries sent a record number of 2,600 athletes to Antwerp.

In many regards, these Games were significant in maintaining and enhancing the weakening Olympic tradition and ideology, which had been tarnished by the First World War. For example, for the VII Olympiad, the Olympic flag, with its intersecting five multicoloured rings (see **Olympic Rings**), and the **Olympic oath** were used for the first time. Another novel feature of the Antwerp Games was the releasing of hundreds of doves during the opening ceremony intended to symbolize the return of peace to Europe. Ironically, the release of the doves was followed by gunshots, somewhat undermining the purpose of the act.

With regard to sporting success, the USA produced the best overall team performance in winning 41 gold medals. However, the real and emerging star of the Games was the 23-year-old Finnish long distance runner, Paavo **Nurmi**, who won medals in the 10,000 metres, the individual cross country race and the cross country team event, and in the 5,000 metres. The Games also featured a week of winter sports, with figure skating and ice-hockey making their Olympic debuts.

Athens Games (1896)

The first modern Olympic Games were held in Athens in 1896. Although Pierre de **Coubertin** wanted Paris to host the first modern Olympics in 1900, the newly founded **International Olympic Committee (IOC)** believed that an earlier date would be more beneficial for the movement. After agreeing on the date, the IOC turned its attention to selecting a host city. De Coubertin's desire for Paris to host the event was again vetoed by the IOC, apparently, based on the arguments of Demetrios **Vikelas** concerning Greece's historical links to the origins of the Olympics, which were sufficiently persuasive to make Athens seem the most appropriate inaugural location.

The IOC's decision to choose Athens as the host city may not have been well thought through, as Greece was experiencing both financial and political troubles at that time. Leading politicians of the country expressed their concern about Greece's social and financial instability and, thus, its capacity to organize and host the Games. The financial demands associated with organizing the Games were frequently discussed. This led the organizing committee to consider resigning. In order to keep the Games alive, de Coubertin and Vikelas made a heroic effort to overturn the defeatist attitude towards the organization of the Games. In search of funding and pro-Olympic sentiment, they approached Crown Prince Constantine who assumed the presidency of the organizing committee in early 1895. With his help and enthusiasm, the organizing committee regained the support of the Greek public, who then provided a significant financial contribution. Nevertheless, more money was still needed and this came from George M. Averoff, a wealthy Egyptian

Greek businessman, who financed the restoration of the Olympic (*Panathi-naiko*) stadium. As a token of appreciation of Averoff's contribution to the first Olympics, a statue was erected in his honour in 1986.

The Games were attended by an estimated 245 athletes, all of them male, from 12 countries. The participation of some of these athletes had been accidental and was simply the result of geographic proximity. Athletes competed in 43 events, including athletics, cycling, swimming, gymnastics, weightlifting, wrestling, fencing, shooting and tennis. A festive atmosphere greeted the participants and a crowd estimated at more than 60,000 attended the opening day of competition. Members of the royal family of Greece played a central role in the support and management of the Games and were regular spectators.

The athletic events were held at the Panathenaic Stadium, the track of which had an unusually elongated shape with sharp turns. The sharpness of the turns made runners adopt a slower, more cumbersome pace. The track and field competition was dominated by athletes from the USA, who won nine of the 12 events. The swimming events were held in the Bay of Zea. Two of the four swimming races were won by Alfréd Hajós of Hungary. Paul Masson of France won three of the six cycling events.

The 1896 Olympics featured the first marathon, which became the highlight of the event and was won by Spyridon Louis, an unknown Greek water car-rier, whose victory earned him lasting admiration and a cup from Michel Bréal, who had been responsible for reviving the marathon event. Although women were not allowed to compete in the first modern Olympics, there was an unofficial female competitor in the marathon. A Greek woman, Stamata **Revithi**, had been refused permission to compete in the men's marathon race and decided to run by herself the following day.

After the closing ceremony, the Greek king awarded prizes to the winners. Unlike today's medal awarding ceremonies, the first place winners received silver medals, an olive branch and a diploma. Athletes in second position received copper medals, a branch of laurel and a diploma. Competitors fin-ishing third did not receive a medal.

The Greek king and many participants supported the idea of holding the next Games in Athens and most of the American competitors signed a letter addressed to the Crown Prince expressing this wish. De Coubertin, however, categorically opposed this idea because he envisaged international rotation as one of the cornerstones of the modern Olympic Games.

Athens Games (1906)

After the initial success of the **Athens Games (1896)**, the **Paris Games (1900)** took place in the shadow of a world fair (*Exposition Universelle Inter-nationale*). The marginalized status of the Games and devaluation of the Olympic spirit concerned the **International Olympic Committee (IOC)**, which proposed the idea of holding interim, or intercalated, quadrennial Games in Athens. The IOC then came up with the following hosting strategy: all of the

Games would be international Olympic Games, but half of them would follow Pierre de **Coubertin**'s idea of having them in different countries, whereas the other half would have a permanent location in Greece. The first intercalated Games were scheduled in 1901 to take place in 1906. The disappointments of the **St Louis Games** in 1904, which again were overshadowed by an exhibition, the Louisiana Purchase Exposition, further damaged the **Olympic Movement** and reinforced the importance of holding interim Games.

These interim Games were successful and helped reinvigorate the Olympic spirit. Interestingly, the 1906 Games are no longer recognized as part of the official Olympic cycle. The personal views of some IOC members caused the organization to retroactively downgrade the 1906 Games, referring to them instead as a 10th anniversary celebration. Some argue that it was de Coubertin himself who later vetoed the results of the 1906 Games and withdrew their official status. In 1948 the IOC executive board, with support from its President, Avery **Brundage** and without discussion, rejected a scholarly petition from another IOC member seeking to reinstate the 1906 Games. In 2003 the IOC executive board once more rejected a coherently argued and well-documented petition from the International Society of Olympic Historians for the reinstatement of the 1906 Games. As in 1948, the matter was not even put to a vote. Hence, today the IOC does not recognize the interim Athens Games as true Olympic Games and does not regard any of its events and achievements, such as new records and medals, as official. This is particularly ironic given that the 1906 Athens intercalated Olympics may well have been the Games that kept the entire movement alive.

Athens Games (2004)

After losing the battle for the right to host the 1996 Olympics, Athens submitted another bid for the 2004 Summer Games. Athens had a relatively comfortable ride, leading all the rounds in the bidding process, and easily defeating the other competing cities, Rome, Stockholm, Cape Town and Buenos Aires. Some 10,625 athletes (4,329 women and 6,296 men) from 201 National Olympic Committees participated in the Athens Games, which consisted of 301 events and were supported by approximately 45,000 volunteers and broadcasted by 21,500 media personnel.

In hosting the 2004 Games, two major issues emerged for the organizing committee. After the attacks of 11 September 2001 in the USA, concerns had grown about terrorism and possible attacks on the Olympics venues. Therefore, Greece increased the budget for security at the Olympics and approximately 70,000 police officers patrolled the city of Athens and other Olympic venues during the Games. Before the Games, Athens also asked for some additional security support from both NATO and the European Union. Fortunately, these security measures were not tested.

The second issue that confronted the local organizing committee was the actual construction of Olympic venues. By late March of 2004, some Olympic

153

building projects were still behind schedule and the pressure to complete them was constantly growing. Consequently, plans had to be slightly modified. For example, a roof, which had initially been proposed as an addition to the Aquatics Centre, was not built. The main Olympic Stadium, the chosen location for the opening and closing ceremonies, was completed only two months before the Games began. Despite international concerns about Greece's schedule for the construction of sites, the host nation finally delivered and by July/August 2004, all venues were completed. Perhaps because of the tight deadline, constant pressure and careless attitude towards safety regulations (according to BBC sources), 14 (mostly migrant) workers died during the construction of Olympic facilities. It was reported that workers did not wear safety gear, were forced to work long shifts and were fired if they complained. Owing to the hazardous construction conditions and their fatal consequences, the 2004 Olympics became known as the 'Greek Tragedy'.

The opening ceremony was spectacular and introduced aspects of Greek mythology and relevant parts of Greek history. However, like other features of these Games, the ceremony was not without controversy. Because the appreciation of human, especially athletic, beauty and bodies were inherent in ancient Greek culture, there was a considerable amount of nudity, most of which was pixellated for the prudish American audience in order to avoid controversy and potential lawsuits.

For the first time, broadcasters were allowed to provide full internet coverage of the Olympics as long as they restricted this service geographically, to protect broadcasting contracts in other areas.

Further indicating the constant and surprisingly high degree of IOC paranoia, the IOC forbade athletes, coaches, support personnel and other officials from setting up specialized blogs or other websites for covering their personal perspectives on and experiences of the Games. They were not allowed to post audio and video messages or photographs that they had taken. An exception was made only if an athlete already had a personal website that had not been set up specifically for the Games.

The sporting highlights of the Games were Michael Phelps's magnificent swimming achievement with six gold and two silver medals, thereby becoming the first athlete to win eight medals in non-boycotted Olympics. Hicham el-Guerrouj won gold in the 1,500 metres and 5,000 metres races and became the first person to accomplish this feat at the Olympics since Paavo **Nurmi**. Argentina achieved a shock victory over the USA in the semi-finals of men's basketball, thus shattering the so-called American 'Dream Team'. These Olympics saw Afghanistan's first appearance at the Games since 1999, having been banned due to the Taliban's oppressive attitudes towards women, and then reinstated in 2002. Paula Radcliffe, the favourite to win the women's marathon, dropped out of the event in an unforgettable fashion. Greek sprinters Konstantinos Kenteris and Ekaterini Thanou withdrew from the Games after allegedly staging a motorcycle accident in order to avoid drug-testing.

Atlanta Games (1996)

The right to host the 1996 Olympics, the 'Centennial Games', was awarded to Atlanta, Georgia, in September 1990, despite competition from Tokyo (Japan), Athens (Greece), Belgrade (former Yugoslavia, now Serbia), Manchester (United Kingdom), Melbourne (Australia) and Toronto (Canada). After the **International Olympic Committee**'s (IOC) decision various rumours came to light that the city had given excessive gifts or, in some instances, used bribery to influence the voting process. Evidence of these practices, however, was to be discovered only after documents were released as part of a 1999 congressional investigation into Olympic **bidding**. In his defence, Atlanta Committee for the Olympic Games (ACOG) Chairman, Billy Payne, said: 'Atlanta's bidding effort included excessive actions, even though processes that today seem inappropriate but, at the time, reflected the prevailing practices in the selection process and an extremely competitive environment'.

The Olympics had a major effect on the landscape of the city as new sports facilities and other venues were built. Park spaces were (re)created and the general transportation infrastructure was improved. During the event, Atlanta received approximately 2m. visitors and an additional 3,500m. watched the events on television. It cost approximately US $1,800m. to stage the Games and the city of Atlanta, heavily reliant on commercial sponsorship and ticket sales, generated a profit of $10m. by the end. However, the extensive involvement of corporate money in the organizing and hosting the Olympics was ambivalently perceived.

As the city of Atlanta was competing with the IOC for advertising and sponsorship capital, the tension between the organizations grew steadily. Broadcast rights negotiations proved to be especially difficult. In order to further increase revenues, much to the disapproval of the IOC, the city licensed street vendors who sold certain products and, thus, facilitated the presence of companies that were not official Olympic sponsors. Dick Pound (*Inside the Olympics*, 2004), former Vice-President of the IOC, observed that the presence of these companies 'led to the curtailing of the Atlanta streets with shabby tents and kiosks that gave the impression of a vast MASH unit'. Interestingly, not many people saw the 'bazaarization' of Atlanta, as 'the head of NBC sport, the US Olympic broadcaster, had a standing rule that any employee who aired a shot of the City of Atlanta ... would be fired on the spot'. The situation was not helped by the Coca-Cola company, which received criticism for ensuring that its products would be the only drinks offered in Olympic venues. Consequently, many argued that the Atlanta Games were the first ones to have been over-commercialized.

Regardless of all the initial finance-driven conflicts, the 1996 Games were given a good start when the cauldron was lit by Muhammad Ali (Cassius **Clay**), semi-paralysed as a consequence of Parkinson's Disease. Unfortunately, the Olympics went downhill thereafter. Despite all the infrastructural improvements, members of the press experienced transportation problems when attempting to get to venues throughout the metropolitan area. Many journalists were late or missed events as some buses broke down and a few drivers got lost. Consequently, there were also delays in reporting the results of events. The situation became even more desperate when the state-of-the-art computer system failed to transfer data from the competitions. This led to missed deadlines and to frustration for journalists from all over the world. In addition, on 27 July, during a concert held in the Centennial Olympic Park, a terrorist tube bomb killed one person and injured a further 110.

Although there were outstanding sporting achievements—for example, Carl Lewis became the third person to win the same individual event four times and the fourth person to earn a ninth gold medal, Naim Suleymanoglu became the first weightlifter to win a third gold medal and Michael Johnson smashed the 200 metres world record to complete a 200 metres and 400 metres double—the Atlanta Games are remembered more for their failings.

In his closing speech, IOC President Juan Antonio **Samaranch** called the Games 'most exceptional', thereby breaking precedent and not saying that they had been 'the best Olympics ever', as he had done at previous Olympic closing ceremonies. This was probably an intentional action triggered by some of the political and commercial tensions between the IOC and AOCOG. Four years later Samaranch returned to his tradition and did call the 2000 **Sydney Games** 'the best ever'.

B

Baillet-Latour, Henri de (1876–1942) (IOC President: 1925–42)

Count Henri de Baillet-Latour, the son of the former governor of Antwerp and senator, Count Ferdinand de Baillet-Latour, was born on 1 March 1876 into a Belgian aristocratic family. He studied arts and philosophy at the University of Louvain, but later specialized in diplomacy. From a young age he was interested in sports, especially equestrianism. In 1903 he began his career as a sport diplomat and was elected as a member of the **International Olympic Committee (IOC)**. Later he founded the Belgian Olympic Committee, of which he became the President in 1923.

After the First World War Baillet-Latour was instrumental in having the Games of the VII Olympiad awarded to Antwerp. Although Belgium had suffered serious social and infrastructural damage in the War and there was only one year to prepare for the **Antwerp Games**, Baillet-Latour shouldered most of the organizing responsibilities and delivered a successful Games.

He gradually strengthened his position within the IOC and after the resignation of Pierre de **Coubertin** he was elected as President in 1925. During his presidency Baillet-Latour devoted himself to maintaining the Olympic ideals and aims. He made extensive efforts to keep sport free from politics and commercialism and to preserve its assumed nobility. His naivety regarding the separation between sports, especially the Olympics, and politics was revealed in 1936 at the **Berlin Games**.

Although the Games of the XI Olympiad were awarded to Berlin in a peaceful climate, when Hitler and the National Socialist party assumed control of the country, the political atmosphere fundamentally changed. Hitler had recognized the value of sports for building a strong and healthy nation and gradually came to appreciate the political opportunities inherent in the modern Olympics. The Nazi leadership decided to harness the international propaganda potential of the Olympics and showcase not only German culture but also Nazi prowess. Years before the Berlin Games, in light of revelations emanating from Germany, the possibility of a boycott was discussed in various countries. Despite his initial misgivings, however, even the President of the US Olympic Committee, Avery **Brundage**, eventually gave his backing to the organizers, claiming that 'politics has no place in sport'. Although Baillet-Latour, too, had many initial concerns about the possibility of the use of the 1936 Games for political propaganda purposes and about discrimination towards ethnic and other minorities in Germany, after paying a visit to Berlin in 1935, his mind was put at ease. He was assured that both visitors and athletes would be given a warm welcome and that they had nothing to fear. Baillet-Latour reassured the international community that Germany had made enormous efforts to separate sport and politics and that there was nothing that should halt the Olympics in Germany. Furthermore, according to him, German athletes were chosen based purely on performance rather than race. After the Berlin Games, he had to admit he had been totally wrong as there were only two Jewish athletes selected for the German team despite the fact that many potential Jewish medallists had been previously identified.

During the entire length of his presidency, Baillet-Latour attempted to keep sports and politics separate. However, his efforts often proved to be futile and unrealistic. He led the IOC until his death in 1942, when he was succeeded by Sigrid **Edström**.

Barcelona Games (1992)

The city of Barcelona, in the Spanish region of Catalonia and birthplace of the then **International Olympic Committee (IOC)** President Juan Antonio **Samaranch**, won the right to host the Games of the XXV Olympiad in competition with Amsterdam, Belgrade, Birmingham (United Kingdom), Brisbane and Paris. In numerous ways, the Games reflected the political changes that had been taking place in various parts of the world as well as changes in the **Olympic Movement** itself. There was no boycott for the first time since 1972. A unified German team competed for the first time since 1964. South Africa returned to the Games after 32 years. The USSR had been dissolved in 1991 resulting in teams from the Baltic states of Estonia, Latvia and Lithuania competing separately and the remaining former Soviet republics making up a 'Unified Team'. Similar arrangements were put in place to accommodate changes in the Balkan region with independent teams representing Slovenia, Croatia and Bosnia-Herzegovina and the remainder of the Yugoslavian athletes competing as 'Independent Olympic Participants'. In terms of the evolution of the Olympic Movement, professionals were allowed to compete officially for the first time and the USA took advantage of this by fielding a so-called 'Dream Team' in men's basketball which included leading National Basketball Association players such as Michael Jordan, Magic Johnson and Larry Bird. The 1992 Games were also politically significant for the host country and the host city. For Spain, it was an opportunity to celebrate and demonstrate the progress made during the immediate post-Francisco Franco era. The regular attendance of the Spanish King Juan Carlos and his family and the better than anticipated performance of the Spanish team ensured that these Games would be seen by the outside world as a Spanish event. On the other hand, this was also an opportunity for the fiercely proud Catalans to promote their 'nation'. The immediate consequence of this was what became known as 'the battle of the flags' with Catalans ensuring that their flag was at least as prominent as that of the Spanish state. The city itself, and particularly its port area, was revitalized by the Games and there ensued a tourist boom which has continued unabated.

Beijing Games (2008)

In 2008 Beijing, the capital city of the People's Republic of China, hosted the Summer Games of the XXIX Olympiad. In so doing, it became the third Asian city, after the **Tokyo Games** and the **Seoul Games**, to host the event. The decision to award the Games to Beijing in the face of competing bids from Toronto, Paris, Istanbul and Osaka was politically controversial. Critics of China's record on **human rights** argued that the country was not yet ready to be given the opportunity to host the Olympics. On the other hand, it was argued by Beijing's supporters that China had already done much to move away from its reputation as a closed, authoritarian society and that having the chance to host the Olympics would provide an additional incentive for it to move even closer to the heart of the global community. Not surprisingly, given its record of public pronouncements on the potentially harmful relationship between sport and politics, the **International Olympic Committee (IOC)** was fearful of the politicization of the Games by both sides. There were also fears of **terrorism** within China in the lead up to and during the Games (although the Chinese Government was able to mount a massive, and often intrusive, **security** operation to prevent such incidents). In the event, all but one of the IOC's members took part in the Games (the only exception being Brunei) and the most politically charged incidents were linked not to the Games themselves but to the **torch relay** that preceded them. The Olympic Flame became a feature of the modern Olympics at the **Amsterdam Games** in 1928. It was not until the **Berlin Games** of 1936, however, that a torch relay preceding the lighting of the Flame in the main stadium took place. In 2008 there were major **protests** against China's record on human rights, together with its treatment of Tibet and its involvement in the war in Darfur, at numerous stages of the torch relay route, accompanied by large-scale security operations (often co-ordinated by China itself) and counter demonstrations by Chinese nationals in a number of countries. During this period, even greater concerns were expressed about the morality of the Games taking place in Beijing than had been voiced when the initial decision was announced. Amongst the leading opposition voices was that of film director, Steven Spielberg, who resigned from his post as artistic adviser to the Games in protest at the role of China in Darfur. The most politically sensitive stage of the torch relay route was to have been that which took the Olympic torch from Viet Nam to Taiwan and onwards to Hong Kong, which has its own 'national' Olympic committee but is regarded, along with Macau, as a Special Administrative Region of China. The Taiwanese Government argued that accepting this route would create the impression that Taiwan was the first stage in the domestic section of the relay and refused to participate. Indeed, related concerns emerged in relation to the organization of the Games themselves with Taiwanese athletes competing as 'Chinese Taipei' as they has done at every games since 1984 and not as 'China Taipei', which would have been the preferred option of China and would have put Taiwan on apparently the same constitutional footing as China Hong Kong. Paradoxically, the Games were held in two separate Olympic committee jurisdictions with the equestrian events taking place in Hong Kong. Impressive new facilities were constructed in Beijing, amongst them the National Stadium which became known as 'The Bird's Nest'. Improvements were made to the city's transport infrastructure and hopes were expressed that the Games would have a positive impact in terms of China's willingness to address environmental concerns. Critics argued, however, that air pollution during the Games had been worse than was officially recognized by China with the effects being camouflaged to some extent by helpful weather conditions. Serbia and Montenegro competed as separate countries for the first time.

Berlin Games (1936)

In 1931 the **International Olympic Committee (IOC)** awarded the Games of the XI Olympiad to Berlin, Germany. Two years later, Adolf Hitler led his Nazi party to power. As the 1936 Games approached, news of the Nazi persecution of Jews and other minorities was already widespread. Jews had not only been excluded from sports clubs, they had been stripped of their civil rights and their German citizenship. Meanwhile, the first concentration camps had been constructed. With the notable exception of a part-Jewish fencer, Helene Mayer, who went on to win a silver medal, Jews were largely removed from the German Olympic team. In light of revelations emanating from Germany, the possibility of a boycott was discussed in various countries, not least the USA. Despite his initial misgivings, however, the President of the US Olympic Committee, Avery **Brundage**, eventually gave his backing to the organizers, claiming that 'politics has no place in sport'. The Republican Government of Spain did boycott the Games, as did the USSR. Arguably the Games are now best remembered for the achievements of the African American athlete, Jesse **Owens**, who won four gold medals. His triumphs are generally regarded as having dealt a serious blow to Hitler's ideological belief in the physical supremacy of the Aryan 'race'. In reality though, the Games represented a massive propaganda success for the Nazis with Germany topping the medal table and other competing nations leaving with the impression that Nazi Germany could not only host an event of this magnitude with considerable panache, but that it was also a peaceable country. There is no doubt that these were the first modern Games at which image mattered as much if not more than athletic performance. The official opening had been preceded by the first ever **torch relay** commencing in Olympia in Greece and the IOC had commissioned Leni Riefenstahl, one of Hitler's personal favourites, to make a film of the Games. In terms of the relationship between politics and sport one of the less well-known incidents at the 1936 Games came in the men's marathon, in which two Korean runners, Sohn Kee-Chung and Nam Sung-Yong won gold and silver medals, respectively. However, because their country was at the time under Japanese colonial rule, they were representing Japan and had been assigned Japanese versions of their names.

The Olympic Stadium is situated in the west of the city and the Olympic grounds have been described as constituting the largest memorial in Berlin. Designed in the 1930s by Werner March, in addition to the stadium itself the site consists of the Bell Tower, Maifeld and Langemarckhalle. Although plans were already afoot to build a stadium, centred around an existing arena designed by March's father, Otto, in order for Berlin to host the 1936 Games, the Nazi seizure of power in 1933 necessitated modifications both in style and ambition. The reconstruction of the stadium was intended to conform to the Nazi regime's preference for neo-classicist architecture. Of special significance in this regard was the Langemarckhalle, a memorial to the German soldiers who had died in the First World War. There is considerable disagreement as to the extent of Hitler's direct influence on the stadium's architecture. Tall stone columns and towers are accompanied by equally monumental sculptures, the *Rosseführer* by Joseph Wackerle and the *Diskuswerfer* by Karl Albiker. Regarded by some as unambiguously Nazi forms, these sculptures actually differ little from those that are to be found in the grounds of Stockholm's Olympic Stadium, built for the 1912 Summer Games, and the nearby Gymnastik-och idrottshögskola in their reflection of a general *Jugendstil* approach to body culture as opposed to fully fledged Nazi ideology.

As for the stadium itself, it has been difficult to separate the structure from its symbolic use in 1936. As a consequence, debates about its future have been suffused with concerns about what should and should not be remembered and about if and how sites of remembrance should be maintained. Now the home of Bundesliga club Hertha BSC, the stadium was central to Berlin's unsuccessful bid to host the Olympics in 2000 and was used during the 1974 World Cup Finals and again, after substantial refurbishment, for the 2006 Finals. One can legitimately argue that it now owes more to the commodification of sport in general and of football in particular, than to the wilder fantasies of Aryanism. To that extent, it might even represent a model for other stadium planners, not least in terms of ease of public transportation facilitated by an adjacent railway station with multiple platforms.

Bidding

Bidding to host the Olympic Games, Summer and Winter, has become virtually a continuous process. There is an initial selection phase in which interested parties are considered in relation to a range of technical requirements. The cities that pass this stage in the process become Official Candidate Cities which are then asked to submit candidature files and are visited by an **International Olympic Committee (IOC)** delegation. Voting takes place seven years prior to the Games for which hosting rights have been sought. The bidding process has been controversial over the years not least when there have been accusations and, indeed, evidence of corruption (for example, the **Sydney Games** in 2000 and the 2002 **Winter Games** in Salt Lake City). Bidding to host the Olympic Games has become something akin to a sporting contest in its own right. Those who run cities and, in most cases, the countries in which they are located, clearly see advantages to be gained from winning the right to host. In some instances, the aim may be to celebrate social and political reconciliation after a period of instability and division (see **London Games 1948**). Another common objective is simply to put the city and the country on the international stage in what is undeniably a highly visible manner. Hosting the Games can be used not only for the purposes of civic boosterism but also as an opportunity for urban regeneration. The decision to bid is not always well received by people who live in the bidding city or elsewhere in the country (for example Nagoya's bid to host the 1998 Summer Games). In addition, as the Games continue to expand in terms of both participation levels and **commercialization**, the prospect of hosting becomes increasingly daunting, making it seem less and less likely that in the foreseeable future, the Summer Games will be hosted by smaller cities such as those which played host in the past—for example **Stockholm**, **Helsinki**, **Antwerp**.

Black September

Black September was the name of a militant Palestinian organization founded in 1970. It became best known for an act of **terrorism** culminating in the murder of 11 Israeli athletes and officials and one German police officer during the 1972 Olympic Games held in Munich, Germany (see **Munich Games**). Using nine of the Israeli team as hostages, the group demanded the release of over 200 political prisoners detained in Israel together with two German prisoners, Andreas Baader and Ulrike Meinhof, the founders of the terrorist Red Army Faction. The Games were suspended for a day and there was a memorial service at which, it was widely felt, the **International Olympic Committee** President Avery **Brundage** devoted more of his speech to the **Olympic Movement** than to those who had died. There was also subsequent criticism of the **security** arrangements in the Olympic village and of the German authorities' rescue efforts. The terrorist attack by Black September prompted tighter security arrangements at subsequent Games.

Boycott

Boycott is a conscious political act; a withdrawal from social and/or commercial relations with an organization or country as a form of protest or punishment. That is, a country may boycott another by voluntarily abstaining from dealing with it, thereby expressing disapproval of the boycotted country's policies.

Boycotts are common in the global sporting arena. The history of the modern Olympics is riddled with them. The first threat of abstention was expressed prior to the **Stockholm Games** in 1912. The Austro-Hungarian monarchy threatened to boycott the Games if Bohemia and Hungary were accepted to participate in the Olympics as independent states. The next threat of abstention was by the USA in 1936 before the **Berlin Games** as a possible response to anti-Semitism in Germany. In both cases, the threat was averted. In 1952, however, the People's Republic of China refused to take part in the **Helsinki Games** because of the participation of Taiwan, leading to the first example of an actual Olympic boycott. In 1956 six countries (Egypt, Iraq and Lebanon; Spain, Holland and Switzerland) withdrew from the **Melbourne Games** in protest at two separate political events: the invasion of the Suez Canal by the United Kingdom, France and Israel, and of Hungary by the USSR. China was also absent, again in response to Taiwan's participation. In 1968 and in 1972 a number of African countries threatened to boycott the **Mexico City Games** and **Munich Games**, respectively, due to the participation of apartheid-ridden South Africa and of Rhodesia (now Zimbabwe). In 1976 several African countries requested the exclusion of New Zealand for continuing to play rugby union with South Africa. This demand was rejected on the grounds that rugby was not an Olympic sport. Consequently, over 20 African countries refused to take part in the **Montréal Games**, creating the first extensive boycott (see also **Gleneagles Agreement**). Four years later, triggered by the USSR's invasion of Afghanistan, the most widespread boycott in the history of the Olympic Games took place for the **Moscow Games** with the USA and 66 other countries refusing to take part. To reciprocate, 13 countries joined the USSR in its decision to abstain from the 1984 **Los Angeles Games**, the third consecutive Summer Olympics for which a boycott was organized. Libya and Iran were also absent for political reasons, but these were more to do with bilateral disputes with the USA. Although the end of the **Cold War** changed the political dynamics of the Olympic Games, there were three absentees from the 1988 **Seoul Games**: Democratic People's Republic of Korea (North Korea), Cuba and Nicaragua. Since then, four Olympic Games have been taken place without any country withdrawing for political reasons. However, the 2008 **Beijing Games** for a while seemed likely to change this trend in the light of numerous **protests** against China's aggressive foreign politics towards Tibet.

The number of boycotts associated with the Olympics is somewhat ironic as one of the original ideas behind the establishment of the modern Games was to create a free international sporting community that no nation-state would manipulate to its political advantage. Clearly, this aim has not been realized so far.

Brundage, Avery (1887–1975) (IOC President: 1952–72)

Born in Detroit, Michigan, on 28 September 1887, Avery Brundage represented the USA in pentathlon and decathlon at the **Stockholm Games** in

1912. In 1929 he became President of the US Olympic Committee during which time he was deeply involved in debates about whether there should be a **boycott** of the 1936 **Berlin Games**. He served as President of the **International Olympic Committee (IOC)** from 1952 to 1972 and, through his position, he sought to resist the growth of professionalism within the **Olympic Movement**. With specific reference to politics, Brundage is best remembered for his role in ensuring that a US team should go to Berlin in 1936. He argued that sport and politics should not mix although many suspected (and continue to claim) that his real motives reflected his empathy with the Nazi regime. Brundage retired from the IOC Presidency in 1972. Away from sport he had built up a successful construction company in Chicago. He was a noted collector of Asian Art and a philanthropist whose financial support was particularly directed towards educational and cultural foundations. A freemason who was a member of the North Shore Lodge No. 937 Chicago, Brundage died on 8 May 1975 and was buried in his adopted home city.

C

Cancelled Olympics

So far there have been three cancelled Olympics: the Berlin Games in 1916, the Tokyo (Helsinki) Games in 1940 and the London Games in 1944. The cancellation of these Games was the consequence of ongoing wars, namely the First World War (1914–18), the Second Sino–Japanese War (1937–45) and the Second World War (1939–45).

For the right to host the Olympics in 1916, three cities had put in a bid. Budapest (Hungary) and Alexandria (Egypt) were unsuccessful and Berlin (Germany) was selected. However, due to the outbreak of the First World War in 1914, these Olympics, which were to be the Games of the IV Olympiad, had to be cancelled. A similar situation emerged in 1940 with the outbreak of the Second World War. Owing to the commencement of the Second Sino–Japanese War in 1937, the **International Olympic Committee** had withdrawn Tokyo's right to host Games of the XII Olympiad and awarded them to the runner-up, Helsinki, Finland. Regardless of changing the venue, after the outbreak of the Second World War, the Helsinki Games, too, had to be cancelled. During the War the Olympics were indefinitely suspended, leading to the cancellation of the Games of the XIII Olympiad which were to be held in London in 1944. The Second World War finally came to an end in 1945 with the collapse of the Axis powers and the surrender of Japan. Subsequently, the **London Games (1948)** took place after a 12-year hiatus.

Centralized doping

Centralized doping was a governmentally organized and supervised medicine-based athletic performance-enhancing practice that emerged during the **Cold War** era in some of the communist countries. The political tension and cultural rivalry of the Cold War filtered into the sport domain whereby sport performance was used to symbolically signify a country's moral and financial superiority over its rivals. Therefore, nations of the two ideological blocs paid careful attention to training their athletes. On both sides, some went as far as administering performance-enhancing chemical substances.

However, in some countries of the communist bloc, doping was handled in a particular way and was highly controlled. The German Democratic Republic (East Germany) was notoriously associated with centrally supervised doping practices. Indeed, recent documentaries and published interviews with former East German athletes have explored the impact on sport of the authoritarian political regimes of the Cold War era. Andreas Krieger—born Heidi Krieger—the former female shot-putter has received specific media attention, being transformed from Heidi to Andreas to a large extent because of the doping regime to which he was exposed. It was also well known since the 1956 Olympic Games that Soviet athletes were on steroids in order to increase their performance. Steroid and its derivatives later became the preferred substance in many of the other communist countries. By the end of the 1960s, on the international level, illegal drug use by athletes had become a common occurrence and the repertoire of performance-enhancing substances began to increase. As a consequence of the growing popularity of medical performance-enhancing practices, the **International Olympic Committee** introduced the first doping control at the **Mexico City Games** in 1968.

Clay, Jr, Cassius Marcellus (Muhammad Ali) (1942–)

Named after his father and a Kentucky anti-slavery campaigner of the 1830s and 1840s, Cassius Clay was born on 17 January 1942 in Louisville, Kentucky, USA. Variously nicknamed 'The Louisville Lip', 'The Champ' and 'The Greatest', Clay won the light heavyweight boxing gold medal at the 1960 **Rome Games**. He claims that he subsequently threw his medal into the Ohio River after having been refused service in a 'whites only' restaurant. As a professional boxer, he won the first of his three world heavyweight titles in 1964 beating Sonny Liston. After this victory, he announced that he had joined the Nation of Islam (the so-called 'Black Muslims') and wished to be known henceforth as Muhammad Ali. In 1966 he attracted both hostility and admiration in almost equal measures by refusing to serve with the American army in Viet Nam, famously quoted as saying: 'I aint got no quarrel with them Viet Cong ... they never called me a nigger'. He was stripped of his title and sentenced to five years' imprisonment. However, in 1970 he was allowed to return to boxing and the following year the Supreme Court reversed his conviction. He regained his world title in 1974 (beating George Foreman) and having lost to another former Olympic Champion Leon Spinks (light heavyweight, Montréal, 1976), in 1978, won the title for a record third time by beating Spinks later the same year. Having survived the controversy of his earlier years, Ali has come to be widely regarded as a great African American and a great American athlete. Undoubtedly one of the most famous sportsmen of all time, he has sought to remain politically and socially active despite having been diagnosed with Parkinson's Disease in 1984. He received a replacement gold medal at the opening ceremony of the 1996 **Atlanta Games**. In 2001 a film of his life starring Will Smith was first screened and he has been honoured in numerous ways including being named as *Sports Illustrated*'s 'Sportsman of the Century' in 1999. The Muhammad Ali Centre in his hometown of Louisville is a cultural and educational facility that seeks

'To preserve and share the legacy and ideals of Muhammad Ali, to promote respect, hope and understanding, and to inspire adults and children to be as great as they can be'. Ali lives in Scottsdale, Arizona, with his fourth wife, Yolanda ('Lonnie').

Cold War (1945–91)

The 'Cold War' was a period of tension and competition that centrally involved the USA and the USSR during the years following the Second World War. Some historians argue that the Cold War grew out of the division of much of Europe into Soviet and American blocs. Although, as wartime allies, the USA and USSR were friendly to one another immediately after the Second World War, their relationship began to deteriorate drastically after the Yalta Conference (4–11 February 1945). In 1948 the USA defended the citizens of West Berlin against a Russian blockade and by 1950 American troops were fighting the North Korean communist satellite of the USSR. During the Cold War era, a serious rivalry developed between the two superpowers, which manifested through military alliances, weapon development, political propaganda, space exploration and sport.

Since the Cold-Warring parties recognized each other's military capabilities, especially after the USSR had developed its own atomic bomb in 1949, having been partially aided by information obtained through espionage, they refrained from traditional forms of war and engaged in clandestine intelligence operations and military actions in strategically selected countries (e.g. Viet Nam 1959–75, Korea 1950–53). One of the most significant international arenas for these two blocs to clash was the Olympic Games. Although, the USSR joined the **International Olympic Committee** and the Olympic 'family' rather late, the achievements of the Soviet athletes soon began to match the political magnitude of their home country. Sporting events between the various countries of the two ideological blocs then began to receive international attention and became high-profile spectacles and symbolic contests between rival cultural and political ideologies.

Comăneci, Nadia Elena (1961–)

The first gymnast to receive the perfect 10.0 for her routine at the Olympics was born in Oneşti, Romania, on 12 November 1961. Comăneci began gymnastics in kindergarten with a local team called 'Flame'. At the age of six she was chosen to attend Béla Károlyi's experimental gymnastics school after he had spotted her and one of her friends performing cartwheels in a schoolyard (a scene in the movie entitled *Nadia* captures well Károlyi's discovery of the young Comăneci). She was one of the first students at the experimental gymnastics school established by Károlyi and his wife, Marta (the couple would later defect to the USA and play a prominent role in American gymnastics). Comăneci showed promise and commitment from an early age and became one of Károlyi's favourite gymnasts. She came 13th in her first Romanian National Championships in 1969, but despite this disappointing result, she persevered and a year later began competing as a member of her hometown team, becoming the youngest gymnast ever to win the Romanian Nationals. In 1971 she participated in her first international competition, a junior competition between Romania and Yugoslavia, winning her first all-around title and contributing to the team gold.

When she turned 12, Comăneci went to live and train at a state-run gymnastics training school. Her parents had initial concerns about letting their daughter move away from the family home at such an early age, but the potential benefits, such as a good education and access to food and travel, proved impossible resist. Comăneci trained with Károlyi eight hours a day, six days a week. For the next few years, she competed as a junior in numerous national contests in Romania and in competitions with neighbouring countries. Her first major international success came at the age of 13, at the 1975 European Championships in Skien, Norway where she won the overall championship and gold medals in every event, except the floor exercise in which she came second. She continued to enjoy success in other competitions in 1975, achieving first place in the all-around, vault, beam and bars at the Romanian National Championships. In March 1976 Comăneci competed in the inaugural American Cup at Madison Square Garden in New York, receiving unprecedented scores of 10 on the vault in both the preliminary and final rounds of competition and winning the overall championship.

Making her Olympic debut at the age of 14, Comăneci became one of the stars of the 1976 **Montréal Games**. During the team competition, her routine on the asymmetric bars was given a 10, the first time in the history of the modern Olympics that a perfect score had ever been awarded to a gymnast. The organizers of the Games were unprepared for this eventuality and scoreboards were not capable of displaying scores of 10. Consequently, Comăneci's perfect marks were reported on the boards as 1.00 instead, which caused some confusion for gymnasts and spectators alike. Comăneci was the first Romanian gymnast to win the all-round title at the Olympics. She also holds the record as the youngest Olympics gymnastics overall champion. Because of the revised age-eligibility requirements in the sport, this achievement cannot currently be legally equalled or surpassed (gymnasts must now turn 16 in the calendar year in order to compete in the Olympics, whereas in 1976 gymnasts had only to be 14 by the first day of the competition). By the time the 1976 Olympics ended, Comăneci had earned seven perfect 10.0s, three gold medals, one bronze, one silver, and massive popularity. She appeared on the covers of *Time*, *Newsweek* and *Sports Illustrated* and returned home to Romania to a heroine's welcome. In spite of the glamour and fame, life after the 1976 Olympic Games was tough for Comăneci. Her parents divorced and Romanian sports officials separated her from Károlyi and made her train with another coach. Upset by this turn of events, Comăneci swallowed bleach to get attention. The government then allowed her to train with Károlyi once again. In 1979 Comăneci allegedly became involved with the son of Romanian dictator, Nicolae Ceauşescu. She has written extensively about these troubled times in her autobiography, *Letters to a Young Gymnast*. Perhaps reflecting the turmoil in her personal life, Comăneci's performance at the 1980 **Moscow Games** did not measure up to her own high standards. Although she won two gold medals and one silver, she fell off the asymmetric bars, considered to be her best event.

She retired from competition in 1984 and became a gymnastics judge and coach to the Romanian national team. Romania's Government still regarded her as a valuable communist asset in sport and fearing that, like others before her, she might defect, they no longer allowed her to travel. Indeed, she was virtually out of sight of the Western press. Moreover, officials of her home country kept her under strict surveillance. Although her sporting achievements gave her a relatively easy life (for instance, she did not have to comply with Ceauşescu's foetus policy), Comăneci eventually could not bear living in a totalitarian regime and, in 1989, fled from Romania through Hungary and Austria and was granted political asylum in the USA. After some initial moving from one place to another, dubious relationships and negative press, she settled in Norman, Oklahoma, and married the American gymnast, Bart Connor in 1996. Comăneci became a US citizen in 2001.

Commercialization

For many years the **International Olympic Committee (IOC)** resisted receiving funding from corporate sponsors. It was not until after 1972 when Juan Antonio **Samaranch** had replaced Avery **Brundage** as President that serious attention began to be paid to the issue of television and advertising rights. Samaranch realized that considerable revenue could be raised from sponsors keen to associate their products with the Olympic brand. Moreover, the need for such revenue streams became all the more apparent with the financial difficulties that beset the 1976 **Montréal Games** and eventually prompted the IOC to appoint marketing expert, Peter Uberroth, to oversee the commercial aspects of the 1984 **Los Angeles Games**. Uberroth succeeded in generating huge profits through TV rights and also by making use of private enterprise for the construction of facilities. The days of the amateur (or amateurish) **Olympic Movement** were over as organizers as well athletes became officially fully professionalized.

The **Seoul Games** of 1988 followed the example set by Los Angeles with leading South Korean companies involved from the bidding process right through to the successful completion of the event. These were also the first Games that had taken place since the formation, in 1985, of the Olympic Partner Programme which in return for sponsorship offered world-wide marketing rights for the Summer and Winter Olympics to partner corporations. In the intervening years, these have included Coca Cola, Samsung, Kodak, Panasonic, Visa and McDonald's.

From modest beginnings, therefore, at least in terms of business acumen, the Olympic Movement has been transformed into a major commercial enterprise as well as an organizer of sporting events. Profits have been generated for the IOC's own use and for the benefit of **host cities**. More generally, however, the Olympics have also played a crucial role in helping numerous transnational corporations to enhance their global image and increase their profits.

Coubertin, Pierre Fredy de (1863–1937) (IOC President: 1896–1925)

The man who revived the Olympics was born on 1 January 1863 into a French aristocratic family. Owing to his family background, he lived the carefree life of the affluent and was involved in boxing, rowing and fencing. He received an education typical of his era and social class and developed an interest in the social sciences. His focus later moved towards education, especially physical education. In order to pursue his ideas about education, he gave up his military career. Between 1880 and 1887, de Coubertin, already an Anglophile, travelled to the United Kingdom and the USA to study those countries' education systems. In the former, he visited Eton College, Rugby and Harrow Schools and the Universities of Oxford and Cambridge. De Coubertin also paid a visit to Much Wenlock in Shropshire, where he encountered the idea of the Olympic Games. Indeed, Much Wenlock is now often claimed to be the birthplace of the modern Olympics. De Coubertin believed that sport, and education through sport, could reinvigorate his nation, which he thought was essential as he, along his fellow French aristocrats, believed that the entire French nation had been demoralized by the Franco–Prussian war (1870–71). His travels reinforced his belief that athletic exercise was of great value for the intellectual development and upbringing of young people.

Seeing the Much Wenlock Games gave de Coubertin the idea that a quadrennially recurring international sporting event would help to propagate his ideas. He used his aristocratic national and international contacts to obtain support for this project and, in 1894, organized and held a conference at the Sorbonne in Paris where the revival of the Olympics was discussed. At this congress the **International Olympic Committee (IOC)** was founded and the principles of Olympism approved. Despite all of these achievements, not everything went according to de Coubertin's plans. For instance, the right to host the first modern Olympic Games was awarded to Athens and Demetrios **Vikelas** was elected to be the first IOC President. After the **Athens Games (1896)**, Vikelas resigned from the presidency, which de Coubertin took over and held for the next 29 years.

During his presidency, de Coubertin made an enormous effort to establish and maintain the **Olympic Movement** and to popularize the Olympics and the ideology with which it is associated. He paid specific attention to the idea of amateurism and the continuation of male hegemony both within sports and in the IOC. De Coubertin was not in favour of women taking part in the Olympics as he believed in the Victorian gender role division and considered women both physically unfit for sport and aesthetically displeasing when taking part in sport. It was, in fact, only after de Coubertin's presidency that women's participation began a slow but steady increase. However, it is to be noted that this increase was not without some difficulties, a case in point being the women's 800 metres debacle at the **Amsterdam Games** in 1928.

De Coubertin was also a prolific writer, producing over 20 books and hundreds of articles during his lifetime. However, some argue that a closer look at these works indicates that most of his writing was fairly repetitive and he had to pay for its publication.

De Coubertin's last public appearance was at the 1936 **Berlin Games** (for his endorsement and appearance Hitler paid him handsomely) where he made a frail attempt to reinforce the argument that partaking was more important than winning, even though the Nazis had a different objective in mind.

Almost completely penniless, de Coubertin died in Geneva, Switzerland on 2 September 1937. In his will, he left directions as to what should happen to his remains. His wishes were carried out and his body was buried in Lausanne, but his heart was removed and buried in Olympia, at the site of the **ancient Olympics**.

Counter Olympics

Whilst the women's sport movement and the Paralympics have increasingly become aligned to the Olympics themselves, attempts have intermittently been made to organize events that demonstrably run counter to the Olympics in terms of their underlying ethos and objectives. The story of socialist sport is fascinating and until relatively recently largely neglected. During the first half of the 20th century, workers' sport organizations operated under the aegis of two rival umbrella organizations, one socialist (the Socialist Worker Sports International) and the other communist (the Red Sport International). National organizations were strong in a number of countries including Germany, France, Austria, Norway and Canada. In all, three Worker Olympiads took place: in Frankfurt am Main (1925), Vienna (1931) and Antwerp (1937). For a time, it appeared that socialist sport was not only possible but that it could actually mount a successful challenge to its bourgeois competitor. So what were its aims and why did its challenge ultimately fail?

Worker sport was intended to provide an egalitarian alternative to bourgeois competitive sport, to commercialism, chauvinism, and the obsession with sporting celebrity and record-breaking. It was concerned with fellowship, solidarity, working-class culture and the pursuit of healthy recreation in an egalitarian atmosphere. Sport was intended to play a significant role in the struggle against nationalism and militarism that pervaded so-called politically disinterested bourgeois sport organizations. If workers were not to become the passive recipients of the sports product, then it was essential that they become involved in the creation of an alternative, counter-hegemonic sporting culture. The approach did not always have revolutionary objectives. In the United Kingdom, for example, the primary concern, was simply a desire to allow workers to play and watch sport in ways that differed markedly from what was currently available.

In some countries, such as Finland, the worker sport movement has retained some of its influence into the 21st century. The worker sport organization (Hapoel) is dominant in Israel. Elsewhere, however, workers' sport organizations have been consigned to the history books along with a variety of other bodies that were formed in the first quarter of the 20th century to promote socialist culture.

There are a number of reasons why the concept of socialist sport failed. A major factor was the constant tension within the worker sport movement between communists and socialists. Indeed, one can argue that specifically social democratic sporting organizations were simply squeezed out in certain parts of Europe by the combined might of the authoritarian practices of communism and fascism. Furthermore, in those societies where parliamentary democracy remained in the ascendant, it was difficult for socialists to compete with the attractions of bourgeois sport. If socialist sport became too competitive and popular, it ran the risk of losing its distinctiveness; if, however, it eschewed bourgeois ways entirely, it inevitably lost support. Undecided as to which path to follow, the worker sport movement in Western Europe eventually succumbed to the hegemony of high-performance, competitive sport. Yet, it managed to leave its mark on the way in which sport has subsequently been organized in some countries, particularly those in the Nordic region.

The Gay Games, established in 1982 in San Francisco by Tom Waddell, represent another attempt to bring athletes and artists together on a basis other than that employed by the **International Olympic Committee (IOC)**. The overall aim of the Games is to encourage inclusion, participation and personal empowerment through sport and other cultural activities. In this way, it is hoped that homosexual men and women can acquire greater self-esteem and be able to enjoy more understanding from non-homosexual communities. The Games, which take place every four years, were intended by Waddell to be called the Gay Olympics. However, the IOC immediately launched a successful legal challenge and the name Gay Games was adopted. Another comparable event, the World Outgames, has taken place twice since 2006 when political manoeuvring in relation to **host cities** prompted what might be called a breakaway movement. Whether the two events can successfully co-exist remains to be seen.

It is worth noting that when the IOC was accused of homophobia on account of its legal challenge to the organizers of the putative 'Gay Olympics', its actions were defended on the basis of the fact that it had sued in the past when the word 'Olympics' was used outside of its auspices. However, the Special Olympics which were launched in Chicago in 1968 did not face a comparable challenge. It was almost certainly felt that to take legal action against an event, the aims of which are to build self-confidence, promote social skills and create a sense of personal achievement amongst people with intellectual disabilities, would attract too much adverse publicity. Seeking its own distinctive approach to competitive sport, the Special Olympics' oath is, 'Let me win. But if I cannot win, let me be brave in the attempt'.

Cultural events

During the lifetime of the **ancient Olympics**, cultural and artistic displays were embedded in the Games as a whole. Pierre de **Coubertin** sought to ensure that this tradition would be maintained in the modern Olympics and from the **Stockholm Games** in 1912 until the **London Games** of 1948, art competitions were held in conjunction with the sporting events with successful artists being awarded medals. Thereafter, cultural events have tended to accompany the main sporting event as opposed to being integral parts of the Games. For example, London boroughs (neighbourhoods) have been given financial support to promote cultural projects in the lead up to the 2012 Games. Critics have argued that this is a poor substitute for the sums of money that have been diverted from the arts to sport in preparation for London 2012. One cultural feature that does remain an essential element of the entire event is the opening ceremony during which hosts produce displays of theatre, dance, music and song which are portrayed as being representative of their nation, whether real or imagined.

D

Diplomacy ('ping-pong')

It is often claimed that sport can make a positive contribution to improving diplomatic relations between previously or potentially hostile states. The phrase normally used to describe this type of diplomatic initiative is 'ping pong diplomacy'. In 1971 the US table tennis team which had been competing in the world championships in Nagoya, Japan, was invited to visit the People's Republic of China. It is widely believed that the visit that followed signified the restoration of diplomatic relations between the USA and China after more than 20 years, although previous diplomatic overtures had helped to pave the way. Rumour has it that the invitation came about after one member of the American team, Glenn Cowan, had been practising with a leading Chinese player and missed the US team bus. Instead, he was given a lift with the Chinese team and their officials. In 1972 President Nixon visited China, thereby consolidating the 'ping pong diplomacy' legend. These events were commemorated at a ceremony in 1981 in the Diaoyutai State Guesthouse in Beijing that was attended by Henry Kissinger, US Secretary of State in 1971, and then in June 2008, at the Richard Nixon Library and Birthplace in Yorba Linda, California. Attempts to follow the example set by 'ping pong diplomacy' by establishing baseball contacts between the USA and Cuba came to nothing, perhaps because, unlike China, Cuba continued to have a good relationship with the USSR, or perhaps because for the Americans to learn table tennis techniques from the Chinese was a rather different proposition from them learning about their 'national pastime' from the communists of a nearby Caribbean island.

Dubin inquiry

Following the drug scandal at the 1988 **Seoul Games**, when the Canadian sprinter Ben **Johnson** tested positive for illegal performance-enhancing substances and was stripped of his gold medal and world record, the Federal Government of Canada established the Commission of Inquiry Into the Use of Drugs and Banned Practices Intended to Increase Athletic Performance. Ontario Appeal Court Chief Justice, Charles Dubin, was appointed to conduct the inquiry, which brought to light some evidence of shocking practices. Having heard several months of startling testimonies about the widespread use of performance-enhancing substances among athletes, Dubin condemned the testing policies and practices of both governmental and amateur sports associations in a report that was released in June 1990. As a direct result, Canada strengthened its drug-testing programme with the creation of the independent non-profit Canadian Anti-Doping Organization in April 1991. This organization is responsible for drug-testing policy, practice and implementation in Canada. The Dubin inquiry also recommended that the **International Olympic Committee** should sponsor the establishment of an independent world doping agency. However, it was not until 1999 that such a body, the World Anti-Doping Association (WADA), was set up with an agreed code, supported by most countries and international federations.

E

Edström, J. Sigrid (1870–1964) (IOC President: 1946–52)

Sigrid Edström was President of the **International Olympic Committee (IOC)** from 1946 until 1952. An accomplished runner in his younger years, he worked with Viktor Balck to bring Swedish sport into a single umbrella organization (*Riksidrottsförbundet*). Sweden's political neutrality no doubt assisted him in his role as a sports leader. He was particularly involved in negotiations concerning the Olympic status of the two Germanys after the Second World War. A bust of Edström by the sculptor Wäinö Aaltonen is to be found in the Swedish city of Västerås.

Environmentalism

An emergent political issue relates to the extent to which sport in general and the Olympics in particular can be more ecologically sound. Fears about global warming and other environmental challenges have prompted serious questions about the future sustainability of international, high-performance sport as presently constituted. As sport has developed, ethical considerations have regularly been set to one side to satisfy the demands of competition and, especially in the case of the Olympic Games, the requirements of **spectacularization**. This is not to suggest that all current sporting practices are harmful to the environment. Indeed, many are only possible if pursued in harmony with nature. Nevertheless, **mega-events** such as the Olympics inevitably create more danger. Although growing concern has led to the emergence of increased general ecological awareness, sport has been relatively slow to react. However, since the 1990s, the IOC has begun to exhibit a greater interest in and awareness of potential environmental problems. This was reflected initially within the context of the **Winter Olympics**, with Lillehammer, Norway, leading the way in 1994 and Nagano in Japan becoming the first Olympic Games to meet stringent environmental criteria. Although Sydney hosted the first Summer Games to be presented as 'green' in 2000, the Winter Olympics have continued to lead the way. The Vancouver Games in 2010 and the Games in Sochi, Russia, in 2014, have both been enthusiastically promoted as 'green'. Perhaps this reflects the degree to which a higher percentage of Winter Olympics events take place in the natural environment and, indeed, in an environment that is arguably most immediately threatened by ecological change. Nevertheless, the bid to host the London Games (2012) placed considerable emphasis on sustainability. Even cynics will accept that the IOC is making many positive statements in this regard, issuing clear instructions to various groupings of sport (land-based, water based, winter, etc.) as to what is expected of them. However, if the Games continue to grow in scale, concerns will remain. Global events such as the Olympics require the movement (by air) of massive numbers of people. 'Greening' the Games themselves cannot obscure that particular environmentally damaging reality.

F

Flags and anthems

Flags and anthems have long been important elements in debates about the politics of **nationalism** in various parts of the world. The **Olympic Movement** has its own anthem and flag, the latter having been designed by de **Coubertin** himself. Consisting of five interlocking rings (in blue, yellow, black, green and red), the flag was raised for the first time in 1920 at the **Antwerp Games**. The Olympic anthem was officially recognized by the **International Olympic Committee (IOC)** in 1958. It is now played during the opening ceremony of each Olympics and also when the Olympic flag is lowered at the closing ceremony. When most spectators and television viewers think of flags and anthems in relation to the Olympics, however, it is likely that their thoughts will turn immediately to those of winning competitors. Since 1924, the national anthems of gold medal winners have been played during the presentation ceremonies whilst the national flags of all three medal winners are hoisted. From time to time, for diplomatic reasons, alternative arrangements are made—for example, during the dissolution of the former USSR or in response to the complex relationship between the People's Republic of China and the Republic of China (Taiwan).

Increasingly, flags have become even more visible at the Olympic Games, with medal winners, draped in their national flags, posing for photo opportunities and taking part in laps of honour. Such displays of national belonging highlight the extent to which the Games remain in essence an international event with competitors performing on behalf of their respective nation-states as well as to satisfy their own personal ambitions.

Freeman, Cathy (1973–)

Born into a poor **aboriginal** family in Mackay, Queensland, Australia, on 16 February 1973, Cathy Freeman won a silver medal in the 400 metres at the 1996 **Atlanta Games** and gold in the same event at the **Sydney Games** four years later. Freeman also lit the Olympic flame at the 2000 Games and there was considerable emphasis on her potential contribution to the politics of cross-cultural reconciliation in Australia. After winning her final, Freeman embarked on a lap of honour carrying both the Australian and the aboriginal flags (despite the IOC's ban on 'unofficial' flags). She had also waved the aboriginal flag at the 1994 Commonwealth Games, and its colours—black (representing the people), yellow (the sun) and red (the land)—featured prominently on her running shoes. Interviewed by the *New York Times* about her athletic career, Freeman, who has experienced racial discrimination throughout her life, commented: 'The time will come when I can be more instrumental in politics and aboriginal affairs. But now I think I'm playing a big part doing what I'm doing'. She retired from competitive athletics in 2003 having been named 'Australian of the Year' in 1998.

G

Gleneagles Agreement

The Gleneagles Agreement was a follow-up to the Singapore Declaration and a direct response to racism, especially as institutionalized through the system of apartheid in South Africa. Member countries of the Commonwealth (a voluntary association of 53 independent countries) affirmed that apartheid in sport, as in other social spheres, is directly counter to the Declaration of Commonwealth Principles as ratified in Singapore on 22 January 1971. The Commonwealth recognized racial prejudice and discrimination as a dangerous, unmitigated vice and pledged to foster **human rights** globally. In 1977, at a meeting at Gleneagles, Scotland, Commonwealth Presidents and Prime Ministers agreed to discourage contact and competition between their sports people and sporting organizations, teams or individuals from South Africa, thereby reinforcing their commitment to oppose racism.

The Commonwealth was an appropriate body to impose a sporting embargo on South Africa because several of the sports with high profiles in South Africa are dominated by Commonwealth member states, including cricket and rugby union. Thus, the Gleneagles Agreement was intended to put pressure on South Africa to change its racist policies. According to the Agreement, Commonwealth countries' representatives were not to visit South Africa to participate in sport. Nevertheless, in 1976, New Zealand accepted an invitation to take part in a rugby union tour to South Africa. This was widely criticized. New Zealand was to endure further international disapproval due to comments made by Prime Minister Robert Muldoon, who claimed that he could not restrict the freedom of his people to travel abroad and that he believed that sport and politics should be kept separate. Consequently, in 1976 several African countries requested the exclusion of New Zealand from the Olympics for continuing to play rugby with South Africa. This proposal was rejected on the grounds that rugby union was not an Olympic sport. As a result, 29 African countries refused to take part in the **Montréal Olympics**, creating the first extensive **boycott**.

Despite the **International Olympic Committee**'s decision to disregard New Zealand's contact with South Africa, Commonwealth countries reaffirmed their full support for the international campaign against apartheid and welcomed the efforts of the United Nations to seek out universally accepted approaches to the question of sporting contacts. They also acknowledged that the full realization of their objectives involved the understanding, full support and active participation of their countries and of their national sporting authorities.

Globalization

Globalization is the term normally used to describe a series of developments which, it has been suggested, have led to a compression of time and space on a world-wide scale. There has been considerable debate as to when globalization began and what its root causes were. For the most part, however, globalization theorists agree that the process operates most markedly in the realms of economics, politics, the media, technology, culture and ideology. Disagreement centres on the implications of these developments. At one extreme, it is argued that globalization will result in the creation of a single global community with a shared culture, thereby obliterating national, regional and local differences (homogeneity). Diametrically opposed to this view is the argument that in response to global forces, people will seek comfort in particularistic traditions and identities (local resistance). Somewhere in between these two extremes is the belief that, whilst difference is undeniably being reduced, variety has been greatly enhanced in most parts of the world.

The Olympic Games are frequently described as a global **mega-event**. From relatively humble beginnings, the **Olympic Movement** has expanded to include most of the world's nation-states. This process itself gives some support to the claim that globalization has its origins in the developed West and represents a form of neo-colonialism whereby certain core countries exercise power over peripheral and semi-peripheral countries, some of which harbour ambitions to join the core at some point in the future. The emphasis in the Olympics on Western sporting activities together with what some would see as the Westernization of traditional Asian sports gives further support to this analysis. In addition, with regard to bidding and hosting the Olympics, most studies indicate that holding the **Tokyo Games** in 1964 and the **Seoul Games** in 1988 was of major significance in terms of assisting Japan and the Republic of Korea (South Korea), respectively, to engage more fully in a Western-dominated global market. **Media coverage** of the Games as well as advertising and sponsorship also tend to be dominated by global corporations which have their origins in the West, particularly in the USA.

It is also important to recognize, however, that the Olympic games, whilst undoubtedly having a global reach, are characterized by international competition. Each **International Olympic Committee** member state has its own National Olympic Committee. Competitors at the Games themselves are members of national teams; they do not take part as individuals regardless of their sporting ability. They enter the arena during the opening ceremony behind their national flags and medal winning is celebrated with ceremonies in which the flags of all three medal winners are unfurled and the national anthem of the gold medallist is played. Furthermore, opening and closing ceremonies, albeit making use of the latest developments in global technology, are enlivened by and arguably best remembered for displays of local, regional and national culture.

It can be argued that no single version of globalization theory is fully supported by the example of the Olympic Games, which certainly have a global impact but which also allow for, indeed encourage, the persistence of distinct national identities. In one crucial respect, however, namely labour **migration**, globalization has clearly impacted on Olympic sport and on international sport more generally.

H

Hägg, Gunder (1918–2004)

Although not an Olympian, Gunder Hägg's career sheds light on the historic relationship between amateurism and the Olympic Games (see **Amateur code**). Born in Albacken, Sweden, on 31 December 1918, Hägg set numerous world record times in middle-distance running during the 1940s when Olympic competition was suspended due to the Second World War, during which Sweden remained officially neutral. In the space of three months in 1942, he set 10 world best times and he held the world record for the mile from 1945 until 1954. Having been branded as a professional in 1946, Hägg was prevented from competing in the 1948 London Olympics. In his defence, he admitted having been given small amounts of appearance money but denied that he had ever asked for payment. Potentially a great Olympian, Hägg was the victim, first, of the impact of global politics on the **Olympic Movement** and, second, of his contravention of the movement's amateur code which was already at best problematic and, at worst, unworkable. Hägg died on 27 November 2004.

Helsinki Games (1952)

The Games of the XV Olympiad were held in Helsinki, Finland, in 1952. The 1940 Games had been awarded to the city but became one of the **cancelled Olympics** because of the Second World War. To win the right to host the Games, Helsinki beat six challengers, five of them from the USA (Detroit, Los Angeles, Chicago, Philadelphia and Minneapolis), together with Amsterdam. Ironically, the USSR competed for the first time at these Olympics in a country that its troops had invaded twice during the Second World War. The Games also saw the first appearance of a team from Israel. West German athletes (from the German Federal Republic and Saarland) competed but there was no team from the German Democratic Republic (East Germany). The Republic of China (Taiwan or, within the context of the **Olympic Movement**, Chinese Taipei) refused to take part in protest against the participation of competitors from the People's Republic of China. During the opening ceremony when flames were lit by two of Finland's most famous athletes, Paavo **Nurmi** and Hannes Kolehmainen, there was a brief disruption when German peace activist Barbara Rotbraut-Pleyer ran to the officials' rostrum to deliver her message of world peace. She was subsequently nick-named 'The Peace Angel'. The Helsinki Olympic Stadium, which had been constructed in the Tölöö district of the city for the cancelled 1940 Games, is still in use. A much-loved landmark, it was the venue for the 2005 World Athletics Championships after various phases of modernization and renovation. The Helsinki Games are sometimes talked about as the last real Olympics—small-scale and conducted in the spirit of sportsmanship with only limited **commercialization**.

Host cities

In many respects the quest to host the Olympic Games has become as com-petitive as the events that take place during the course of the event. This is hardly surprising given the increasing use of sport and sporting venues for the purposes of civic boosterism. In the USA and Europe the desire to boost the image and reputation of a city through sport can be partly satisfied by the construction of major **stadiums** and arenas (for football in the case of Europe, and for baseball, American football, basketball and ice-hockey in North America), and the capacity to attract professional franchises (in the case of North America). A glance at the history of the Olympics, however, reveals that European and North American cities have also hosted the vast majority of Summer Olympics. In light of the growing financial and security risks involved, one is entitled to ask the question: 'why?'.

For many civic leaders, hosting the Summer Games is testimony to the global world status of their city. In the lead up to 2102, for example, there can be no doubt that much will be made of the fact that London will be hosting the Games for the third time. The message for Londoners is clear. You can be proud of your city because once again it has provided evidence of its standing in the world. More widely, the choice of London underlines its position as capital city. This latter point is significant not least in the case of cities such as Montréal and Barcelona (arguably Barcelona gained more than other host cities in terms of increasing tourism as a direct consequence of the urban regeneration that took place in order to host the Games), which have hosted the Games despite not being the capital cities of their respective nation-states. What these cities were able to offer, however, was an element of sub-national or quasi-national distinctiveness, which neither Manchester nor Birmingham were able to demonstrate when seeking to join London as a British Olympic host city. Whilst neither Melbourne or Sydney can claim sub-national dis-tinctiveness, hosting the Games was clearly regarded by many in both cities as evidence of status within Australia.

Another factor in the decision to award hosting status relates to global events. For example, both Berlin and Tokyo were allowed to host the Games as part of the process whereby they were brought back into the international fold following the First World War and Second World War, respectively. In addition, Tokyo became the first Asian city to play host to the Olympics, thereby confirming its global standing. Further evidence of this type of transition was provided when Seoul and then Beijing were given the right to host the Games. Over the years the idea of a permanent site for the modern Olympics—probably in Greece—has been mooted at regular intervals. Whether this would be acceptable to the Greek Olympic Committee or feasible in relation to cost are open to some doubt. In any case, there appears to be a never-ending queue of cities which have not hosted the Games but are keen to do so, or have been hosts and are willing to be so again. However, in addition to the financial implications—almost ruinous in the case of Montréal—which continue to present difficulties for cities in the developing world, the global terrorist threat cannot be overlooked. Indeed, it may increasingly come to be recognized that the Olympic Games can only take place in cities where democratically elected politicians are willing to set aside basic civil liberties in order to minimize the likelihood of politically motivated violence or, as in the case of Beijing, in countries with a centralized (and highly militarized) system which offers fewer civil liberties in the first instance.

Human rights

The concept of human rights refers to the basic rights and freedoms to which all humans should be entitled. Examples of human rights may include civil and political rights, including the right to life and liberty, freedom of expression and equality before the law; and social, cultural and economic rights, including the right to participate in religious and cultural activities, the right to work and the right to education.

When we consider the fundamental principles of Olympism as outlined in the **Olympic Charter**, links between human rights and Olympic principles can be easily identified (see page 11 of the Olympic Charter). For instance, the Olympic Charter states that: 'Any form of discrimination with regard to a country or a person on grounds of race, religion, politics, gender or otherwise is incompatible with belonging to the Olympic Movement'. In reality, if the **International Olympic Committee (IOC)** had adhered to these principles, the 1936 **Berlin Games** would not have taken place and, indeed, the USA (and some other Western countries) should have been excluded from the Olympic Movement during much of the 20th century. Furthermore, when giving the 'Black Power' salute at the 1968 **Mexico City Games**, the athletes should not have been reprimanded and expelled from the Olympic village. On the contrary, Tommie Smith and John Carlos should have been perceived as athletes standing up for their rights in the name of human dignity as outlined in the Olympic Charter, which condemns the evil of racial and other forms of discrimination.

Another human rights-related statement in the Charter outlines that: 'The practice of sport is a human right. Every individual must have the possibility of practising sport, without discrimination of any kind and in the Olympic spirit, which requires mutual understanding with a spirit of friendship, solidarity and fair play'. After reading this statement, one wonders whether the IOC considered the rights of those migrant workers who died during the construction of the venues for the 2004 Athens Olympics or whether the unfair treatment of those people who were forced to vacate their houses for the sake of building Olympic venues in Beijing were taken into account. Unfortunately, as history indicates, the IOC hardly ever concerns itself with the sufferings of minority and/or disadvantaged populations. (It is noteworthy that all the IOC presidents have come from aristocratic/wealthy families.) It celebrates principles that can be showcased in the well-orchestrated sporting arena of the Olympic games, where the glamour and glitter of athletic achievements and perfectly sculpted athletic bodies outshine the personal troubles of those who have been exploited, abused and dehumanized for the sake of the largest sporting event on the planet. The examples mentioned above are only a selection from a long list of human rights violations that have been carried out in the name of, during or alongside the Olympics.

I

Illegal performance enhancement

Illegal performance-enhancing substances and procedures, e.g. doping, have a long-standing and controversial history in sport in general and in the Olympics in particular. It would be inaccurate to say that the use of performance-enhancing chemicals has always been frowned upon and banned. Although nowadays, especially since the foundation of the World Anti-Doping Association (WADA) in 1999, the **International Olympic Committee (IOC)** presents a united front when it comes to drug-taking, there is evidence to suggest that in the early years of the modern Olympics, athletes were known to take performance-enhancing aids such as heroin, cocaine, caffeine and strychnine mixed with brandy. While there was always a moderate level of drug-taking, the **Cold War**-induced rivalry fundamentally changed the tapestry of sport-related performance enhancement. The athletes of the USSR and the communist bloc in general received, often through **centralized doping**, testosterone and its derivatives to enhance their athletic achievements and to prove their countries' ideological and socio-cultural superiority. This system seemed to function well and athletes of the Eastern communist bloc began to score highly on the medal table. Inevitably, the USA, as concerned losers, introduced a medical counter measure by administering steroids to their athletes. For the sake of proving and gaining ideological superiority, athletes' health and bodies were maimed and sacrificed on the altar of high-level performance. The use, and abuse, of drugs has led to a number of tragic incidents at international competitions with athletes fainting, collapsing and even dying due to the combination of drugs and gruelling training regimes. Consequently, in order to preserve the 'pureness' of sport (not necessarily to save athletes' lives), the European Council founded a Committee on Drugs, the success of which in refraining athletes from using drugs was minimal. However, following the drug scandal at the **Seoul Games**, when Ben **Johnson** tested positive for illegal performance-enhancing substances and was stripped of his gold medal and world record, the Federal Government of Canada independently established the Commission of Inquiry Into the Use of Drugs and Banned Practices Intended to Increase Athletic Performance. This Commission then launched the so-called **Dubin inquiry**, which brought to light evidence of shocking practices. Subsequently, in 1998, the Festina scandal gave the final boost to the IOC and the world of sport to establish and, more importantly, fund an organization that would solely be concerned with checking and detecting the illegal use of performance-enhancing substances and practices within the realm of professional sports. Although the establishment of WADA has altered drug-taking practices, with the invention and application of ultramodern designer drugs (created and marketed so as to evade existing drug laws), the battle continues.

Imperialism

For many reasons, the Olympic Games can be said to represent the triumph of the West. Established in Europe and inspired by the Games of ancient Greece, the modern **Olympic Movement** has over time extended its sphere of its influence to almost every part of the world. Despite this global reach, however, both the Summer and the **Winter Games** have for the most part been contested in Europe and North America. The Olympics have never taken place on the African continent whilst in 2016, Rio de Janeiro will become the first South American host city. Only three Asian cities (Tokyo, Seoul and Beijing) have hosted the Summer Games. In addition, the overwhelming majority of sports that make up the Games' programme originated in the West. Opponents of the argument that the Olympics provide clear evidence of unidirectional cultural imperialism will point to the fact that both judo and taekwondo, martial arts originating in Japan and Korea, respectively, are part of the Olympic programme. Judo was first introduced to the Olympics for the **Tokyo Games** of 1964. There was no judo at the 1968 Games but the sport has been part of the programme of events at every Games since, with female judo players first being awarded medals at the 1992 **Barcelona Games**. Japan has dominated the judo competition, winning 65 medals in total, with France and the Republic of Korea (South Korea) winning 37 medals each. Taekwondo made its first appearance at the Olympics (as a guest sport) in 1988 at the **Seoul Games**. It was first contested as a medal event at the 2000 **Sydney Games** and has remained on the programme since then. South Korea has been the most successful Olympic taekwondo country, having won 12 medals to date. Advocates of the cultural imperialism thesis would argue that although these Asian sports have been incorporated into the Olympics, this has come at a price. In particular, traditionalists claim that judo has become Westernized in the process with increasingly less attention being paid to the non-competitive elements of each martial art so that even the Japanese have been forced to adopt a more Western approach towards judo. It must be recognized, however, that even most Asian supporters of these sports take great pride in their inclusion in the Olympics. Furthermore, in broader terms, there is a strong case for arguing that both Japan and South Korea (and almost certainly the People's Republic of China in the future) gained much from hosting the Games in terms of international recognition and self-confidence.

International Olympic Committee (IOC)

As part of the ambition of Pierre de **Coubertin** to revive the Olympic Games, the IOC was established at the International Athletic Congress in Paris in 1894. The first IOC President was Demetrios **Vikelas** who was instrumental in making the decision regarding the location of the first Olympic Games. Even though de Coubertin preferred Paris to Athens as the first host city of the modern Olympics, Vikelas's argument concerning Greece's historical links to the origins of the Olympics persuaded IOC members that Athens would be the most appropriate venue. After the **Athens Games (1896)**, de Coubertin became the President of the IOC, a position he held for 29 years, having a lasting impact on the **Olympic Movement**. While the IOC originally consisted of 14 men, mainly chosen by de Coubertin himself, currently the number stands at 111, reflecting the increased global significance of both the Games and the IOC. The official languages of the IOC are French and English. However, at IOC sessions, simultaneous translation is also provided into German, Spanish, Russian and Arabic.

The IOC is the main ruling body of the Olympic Games and owns all the tangible and intangible properties associated with them. For example, the IOC is the sole owner of the Olympic logos and has sole responsibility for the Olympic flag, the motto, the creed and the anthem (see **Flags and anthems**). This demonstrates a strong commercial interest on the part of the IOC. Indeed, the revenue generated by the Olympic Movement is carefully managed. The IOC manages broadcast partnerships and The Olympic Partner Programme (TOP), which is a world-wide sponsorship programme. The Organizing Committees for the Olympic Games (OCOGs) manage domestic sponsorship, ticketing and licensing programmes within the host country under the direction of the IOC. According to the IOC's own figures, the Olympic Movement generated a total of US \$2,570m. from broadcast rights and \$866m. from TOP during the most recent Olympic quadrennium (2005–08). The IOC distributes approximately 92% of Olympic marketing revenue to organizations throughout the Olympic Movement to support the staging of the Olympic Games and to promote the world-wide development of sports and the Olympic ideology. It retains approximately 8% of Olympic marketing revenue for the operational and administrative costs of governing the Olympic Movement.

Initially, the IOC headquarters were in Paris but they were moved to Lausanne, Switzerland, in 1918 and have been located there ever since. The IOC has had eight presidents since its inception. With the exception of Vikelas, these presidents have had a tendency to be long serving (they are elected for a minimum of eight years, with the possibility of being re-elected for another four). IOC presidents have had fundamental effects on the development of the organization and of the Olympic Movement more generally. One of their recurring agendas has been to keep politics out of the Games and out of sports. This aim is controversial, especially when one considers the content of the **Olympic Charter**, which outlines that the IOC is also responsible for promoting specific socio-cultural values and practices relating to sport, in particular, and to wider social and political issues, in general. Furthermore, the IOC is a globally significant, powerful organization with close, bilateral links with over 200 National Olympic Committees and over 200 International Sport Federations. Through these links, the IOC, and especially its president, possess the power to exert a far-reaching influence on the development of modern sports and, by extension, modern culture.

For example, the IOC is in a position to contribute to the preservation of male hegemonic dominance both within the Olympics and sport in general. Indeed, until the early 1980s, the IOC was an exclusively male preserve. The first two female members were admitted in 1981 and, although the number of women involved in the IOC has been growing, the organization remains a predominantly male preserve.

J

Johnson, Benjamin Sinclair ('Ben') (1961–)

Ben Johnson was born in Falmouth, Jamaica, on 30 December 1961. He emigrated to Canada (Scarborough, Ontario) in 1976 where he met Charlie Francis, the Canadian national sprint coach, and joined the Scarborough Optimists Athletic Club. Johnson was a promising athlete, his first international success coming when he won two silver medals at the 1982 Commonwealth Games in Brisbane, Australia. He finished behind Allan Wells of Scotland in the 100 metres with a time of 10.05 seconds and was a member of the Canadian 4x100 metres relay team. The 1983 World Championships in Helsinki did not bring him further success as he was eliminated in the semifinals, finishing sixth with a time of 10.44.

Johnson's international athletic career took off, however, in 1984 when, at the **Los Angeles Games**, he reached the 100 metres final. After a false start, which may have been an attempt to break American Carl Lewis's concentration (the relationship between the two men was not without tension), he won the bronze medal behind Lewis and Sam Graddy with a time of 10.22. He also won a bronze medal with the Canadian 4x100 metres relay team. By the end of the 1984 season, Johnson had established himself as Canada's top sprinter and on 22 August in Zurich, Switzerland, he broke the Canadian 100 metres record with a time of 10.12 seconds.

Initially, Johnson regularly lost to Carl Lewis and, for a short while, it seemed that he had reached the peak of his athletic career. Defying this belief, however, his performances began to improve noticeably in 1985 and, after seven consecutive losses, Johnson finally beat Lewis. In 1986, at the Goodwill Games, he beat his rival again, breaking the 10 seconds barrier in doing so with a time of 9.95 seconds. By the time of the 1987 World Championships in Rome, Johnson had won his four previous races against Lewis and had established himself as the world's best 100 metres sprinter and a Canadian national hero. In Rome, Johnson gained world fame by beating Lewis for the title and setting a new world record of 9.83 seconds. After breaking the world record, he became a celebrity and his popularity in Canada grew even more. Indeed, in 1986 he was awarded the Lou Marsh Trophy as Canada's top athlete.

Both the peak and the beginning of the end of Johnson's athletic career came with the 1988 **Seoul Games** at which he beat Carl Lewis and broke his own world record in the 100 metres by 0.04 seconds. Canadians basked in the athletic achievement and glory of Ben Johnson and national newspapers covered the occasion by concocting words such as 'Benfastic' (*Toronto Star*, 25 September 1988). Unfortunately for Johnson and for Canada, the celebrations proved to be short-lived. Three days after the final, an illegal performance-enhancing substance (Stanozolol) was detected in the Johnson's urine sample. Consequently he was disqualified, stripped of both his gold medal and his world record, was obliged to leave the Olympic village and received a two-year ban from athletic competition. Within the space of one day, his international and national popularity dramatically declined and his status shifted from hero to villain (or as the title of his biography puts it, 'Hero to Zero'). Whilst on the top, breaking records and winning medals, Johnson, despite his Jamaican origins, was considered a true Canadian hero, a title he lost in a split second after his wrong-doings had come to light. The Canadian media immediately labelled Johnson a cheat and the Canadian Prime Minister, Brian Mulroney, stated that: 'It was … a tragedy for Johnson and a great sadness for all Canadians.' Initially, Johnson denied the accusations but, in 1989, at the **Dubin inquiry** he confessed that he had cheated. To the astonishment of many, his coach admitted to the inquiry that Johnson had been using steroids since 1981.

After his ban in 1991, Johnson attempted an athletic come back without much success. He only reached the semi-finals of the 100 metres at the 1992 **Barcelona Games**. In fact, things went from bad to worse for Johnson when in 1993, he was found guilty of doping at a race in Montréal (this time for excess testosterone) and was subsequently banned for life by the International Association of Athletics Federations (IAAF). Federal amateur sports minister, Pierre Cadieux, called Johnson a national disgrace and suggested he consider moving back to Jamaica.

After receiving his life-long ban from athletics, Johnson ran charity races and become a coach. In June 2005 he launched the 'Ben Johnson Collection' in Canada. The 'Collection' consists of sportswear and casual clothing for both men and women. In March 2006 a television advertisement for an energy drink, called 'Cheetah Power Surge', featuring Johnson, was aired. The advertising campaign was built on Johnson's controversial sporting background, encouraging the viewers to try 'Cheetah'. Not surprisingly, it received a mixed public reception. Today, Johnson spends much of his time with his daughter and granddaughter. He also continues to coach.

K

Killanin, Michael Morris (3rd Baron Killanin) (1914–99) (IOC President: 1972–80)

Lord Killanin served as **International Olympic Committee (IOC)** President from 1972–80. Born on 10 July 1914, in 1927 he succeeded his uncle as Baron Killanin, an English peerage, thereby entitling him to take his seat in the British House of Lords when he reached the age of 21. He worked as a journalist before seeing military action during the Second World War. In 1950 Killanin became head of the Olympic Council of Ireland. He was made his country's representative on the IOC in 1952, becoming the IOC's Senior Vice-President in 1968 and succeeding Avery **Brundage** as President in 1972 in the wake of the terrorist attack which had marred the **Munich Games**. During his presidency, Killanin had to deal with a number of political and other problems. Not only were the **Montréal Games** a financial disaster, they were also boycotted (see **Boycott**) by a number of countries protesting against the participation of New Zealand owing to that country's ongoing sporting relationship with the banned South Africa. Even more serious, perhaps, was the boycott of the **Moscow Games** by the USA and some of its allies. Killanin was bitterly opposed to this practice of linking politics to Olympic participation and non-participation. Indeed, so critical was he of the use of the Games for propagandist purposes that he even sought to end the use of national uniforms, **flags and anthems**. He was also keen to relax some of the rules governing amateurism which had been so vehemently defended by his predecessor, Brundage (see **Amateur code**). After the Moscow Games, Killanin was replaced as IOC President by Juan Antonio **Samaranch**. Away from sport, he was active in film and theatre production. He was also Ireland's Honorary Consul-General of Monaco from 1961 until 1984. He died on 25 April 1999 and was buried in Galway, Ireland.

Korbut, Valentinovna Olga (1955–)

The 'little girl' with an elfin figure, Olga Korbut, the so-called 'Sparrow from Minsk', who fundamentally changed women's gymnastics was born on 16 May 1955 in Grodno, Belarus—part of the USSR at that time. Korbut, an active child, began her gymnastics training at the age of eight in a Belarusian sport school but, as her talent in gymnastics became obvious, she was transferred to the group led by Renald Knysh a year later. Under the watchful eyes of Knysh, Korbut's gymnastic career took off and she began to establish a national reputation. In 1967 Korbut entered the Belarusian junior gymnastics championships and a year later she won gold medals in the vault, balance beam and asymmetric bars at the *Spartakiade*. In 1969, for her first Soviet national championship, the age rules had been altered to allow the 14-year-old Korbut to participate. At this competition, she performed two new elements that shocked the audience: 'the Korbut Salto' (a backwards-aerial somersault, launching and then landing on the balance beam) and 'the Korbut Flip' (while standing on the high bar of the uneven bars, performing a back flip into open air and catching the bar on the way down). Her astounding new moves triggered ambivalent reactions. Whilst most people admired her athletic talent and unorthodox routines, some officials complained that the new elements were not in keeping with traditional gymnastics. Consequently, she achieved only fifth place in her first major competition. Mostly because of controversy around her gymnastic style and relatively young age, she was put in the reserve team for a few years. None the less, in 1971 she achieved fourth place in the Soviet national championships and earned the Master of Sports title—a prestigious award granted to those who excelled in sports.

Korbut's international fame arrived at the 1972 **Munich Games** when she made gymnastic history, impressing both the audience and judges and winning three gold medals and a silver. Following the Olympics, Korbut attracted extensive media and public attention and toured the USA and Europe. During this time she visited US President Richard Nixon and met the Prime Minister and the Queen of the United Kingdom. Olga Korbut's official website (www.olgakorbut.com) claims that: 'she did more to ease the tensions of the **Cold War** than all the politicians and diplomats of the day put together'. Although Olga did indeed capture the heart of the international public and travel to several Western countries, the above statement is somewhat naive.

Korbut attended the **Montréal Games** in 1976 and she again was part of the winning Soviet team. Moreover, she took an individual silver medal on the balance beam. However, her individual career as a gymnast began to decline due to the appearance of a young gymnast, Nadia **Comăneci**, who scored the first perfect 10.0 in Olympic history.

In 1977 Korbut officially retired from gymnastics, completed her studies and accepted a coaching job in Minsk, Belarus. In 1978 she married a Belarusian musician, Leonid Bortkevich, and a year later gave birth to her only son, Richard. In 1986, when the Chernobyl disaster struck, spreading a radioactive cloud over almost the entire Eastern Europe, people in that region, especially those living close to the epicentre of the nuclear blast, began to panic. Being afraid of the affects of the radiation, Korbut and her husband sent Richard to New Jersey in the USA to live with relatives. In 1991 they followed their son, emigrating to the USA and settling in Atlanta, Georgia. Korbut's initial years in the USA were not without problems. In 2000 she went through a divorce and a year later the police retrieved serious, incriminating evidence from her house. Although she had not lived there for over two years, her reputation dropped to an all-time low. Unfortunately for Korbut, her troubles were still not over. In 2002 she was arrested for shoplifting (US $19-worth of food), but avoided more serious punishment by paying a fine. After experiencing such chaos in her life, Korbut resurrected her coaching career and took up a job in Scottsdale, Arizona, where she still gives private lessons to gymnasts and dancers.

L

London Games (1908)

From relatively inauspicious beginnings, the London 1908 Games have come to be widely recognized as the first modern Olympics in any meaningful sense. The Games of the IV Olympiad were initially intended to be held in Rome. However, the eruption of Mount Vesuvius in 1906 meant that funds had to be diverted to disaster relief for the city of Naples. This meant that London had only 18 months to prepare for the 1908 Games—a remarkably short time when one considers how long host nations now have to ready themselves. Nevertheless, the White City Stadium was hastily constructed and the Games went ahead on time, owing in no small measure to the organizational abilities of Baron Desborough of Taplow.

The Games spanned the period from 27 April to 31 October. Irish athletes refused to take part at all in protest against the British Government's refusal to grant independence to Ireland. However, most of the political debate prompted by the Games centred on the issue of flags. For the first time at the Olympics, athletes were organized into national teams and were required to march into the stadium for the opening ceremony behind their national flags. The Finnish competitors objected to being asked to march behind the flag of their country's current occupiers, Russia, and did not take part. In addition, both the Swedish and the American delegations had been angered by the fact that their flags were not on display at the stadium itself and, in protest, the Swedes refused to participate in the opening ceremony. Controversy and confusion still surrounds the American response. It is believed, however, that the US flag bearer, Martin Sheridan, refused to dip his flag in front of the royal box. The political nature of this protest has been amplified by suggestions that it was linked to anti-British sentiments within the Irish-American community to which Sheridan belonged, as did the US team manager, James E. Sullivan, who became something of a hate figure in the eyes of the British public. On the other hand, alternative versions of these events have questioned if, in fact, Sheridan was even the US flag bearer at the ceremony.

During the ceremony, the Olympic creed was publicly proclaimed for the first time. Gold, silver and bronze medals were awarded for the first time at these Games and in response to a number of controversial decisions, the **International Olympic Committee** decided to standardize rules for track and field events and to draw future judges from an international pool rather than exclusively from the host nation.

In terms of Olympic history and the future of the Games, arguably one of the most important outcomes of the 1908 Games was the fixing of the length of the modern marathon at 26 miles and 385 yards, the exact distance between the British royal family's residence, Windsor Castle, and the royal box at the White City. The race itself is remembered because of the fate that befell the unfortunate Italian athlete, Dorando Pietri, a baker from Carpi, who entered the stadium in first place but collapsed on a number of occasions, went in the wrong direction at one stage and was eventually helped across the finishing line. Pietri was subsequently disqualified and the gold medal went to Johny Hayes of the USA. In recognition of his bravery, Pietri was presented with a consolation gold cup by Queen Alexandra and with £300 by Sir Arthur Conan Doyle, creator of the fictional detective, Sherlock Holmes. Pietri turned professional and twice beat Hayes in the USA. He earned enough money to be able to retire to Italy in 1911 and buy a hotel in San Remo. Another casualty of the marathon was the Canadian aborigine, Tom Longboat, one of the pre-race favourites, who collapsed after 19 miles, rumoured to have been given champagne by his supporters.

London Games (1948)

The London Games of 1948 have also become known as the 'austerity Olympics' because of the economically challenging post-Second World War conditions in which they took place. Great Britain was still in a state of slow recuperation, which was not aided by the new Labour Government spending a large amount of money on funding both domestic and foreign military operations. Food and other staple products, from eggs to petrol, were strictly rationed. It is, therefore, little surprise that there were controversies concerning the organization of the 1948 Games.

Because of the grim socio-economic conditions of the time, the press chiefly adopted a defeatist attitude and questioned both the relevance of the modern Olympics and the likely success of London as a host city. It appeared that the critique from the press was not completely speculative and the organization of the Games was under-funded and progress was slow. In early 1948 Denzil Batchelor, a journalist, made a phone enquiry regarding the ongoing preparations for the Games at the Ministry of Works. After a long hold up and having being passed from one department to another, he was told that there was no planning underway. Although the organizers did eventually begin to develop and execute plans, a great many preparatory works were left to the last few weeks. For example, just two weeks before the opening ceremony, the greyhound track at Wembley was dug up and filled with 800 tons of cinder. Placing signs and flags throughout London indicating Olympic venues and welcoming international visitors was also a last minute job.

Wembley Stadium was selected as the main Olympic venue. However, along with other sporting facilities, Wembley required fundamental repairs, the cost of which the government could not fully bear. Consequently private help was sought. Sir Arthur Elvin (the Governor of Wembley) managed to persuade Wembley Stadium Ltd to finance most of the necessary reconstruction. Unfortunately, domestic support could not fill all the financial gaps and various donations were received from some of the competing countries. For example, Sweden and Finland both provided timber free of charge to replace the rotting floors of sporting venues. Moreover, the Canadian swimming association supplied two Douglas Fir trees to make diving boards. Even prisoners of war were drafted into building new Olympic facilities.

In addition to domestic organizational difficulties, the London organizing committee was facing other problems as well. Invitations to potential participating countries and their replies were often delayed due to the ineffective international mail service. Furthermore, like the United Kingdom, many of the potential participant countries were still faced with the hardships of post-war conditions and, thus, were slow and inefficient in letting the organizers know about key details such as arrival times and numbers of athletes. Therefore, the organizers had to deal with a plethora of late and inappropriate entries, and schedule events without knowing the exact number of athletes that would be taking part in them.

There was also concern about the quality of competition. First, a large number of the outstanding athletes of the pre-war era had died or been seriously injured in the war, leaving a huge sporting vacuum. Second, athletes simply had not had sufficient time and resources to prepare for the London Games. Since food was still rationed (e.g. meat rations had just been further reduced before the Olympics), organizers struggled with feeding both domestic and foreign participants. Although the government did not issue extra rations of foodstuff for British athletes, foreign competitors were allowed to bring into the country up to 25 pounds of food. In spite of the lack of governmental support, most of the British athletes had a rich diet as they were sent food parcels from anonymous supporters. Restaurants frequently served extra helpings to members of the British squad and international athletes often shared their food allowance with British ones. This created a unique atmosphere and gave a flavour of international friendship to the Games, which turned out to be more successful than many had feared.

Los Angeles Games (1932)

In 1932 Los Angeles hosted the Games of the X Olympiad. With the world in the midst of the Great Depression, there had been no other bidders and Los Angeles itself welcomed fewer than half the number of competitors than had competed in the **Amsterdam Games** of 1928. The Games were officially opened by the US Vice-President, Charles Curtis, instead of President Herbert Hoover, who did not attend at all, the first head of a host nation's government not to do so. The Games lasted for 16 days, thereby establishing the time period for competition which was to become the norm henceforward. Another first was the use of a victory podium with the raising of medal winners' flags and the playing of gold medallists' national anthems. It has also been suggested that Los Angeles was responsible for the first official Olympic village, catering for male competitors only, although there had been embryonic villages since the London Games of 1908. The Los Angeles Olympic Stadium—officially the Los Angeles Memorial Coliseum—was built during the early 1920s as a memorial to the dead of the First World War. It was subsequently used for the opening and closing ceremonies for the 1984 Games and for the track and field events. Absent from the 1932 Games was the great Finnish runner and celebrated Olympian, Paavo **Nurmi**, who had been found guilty of flouting the IOC's rules on amateurism. A gold medal in swimming was won by Clarence 'Buster' Crabbe who later achieved even greater fame as a film star playing the role of Tarzan, Buck Rodgers and Flash Gordon. As a harbinger of what lay ahead in terms of world and Olympic politics, the Italian winner of the 1,500 metres track event, Luigi Beccali, gave the fascist salute on the victory podium.

Los Angeles Games (1984)

The bidding contest for the right to host the 1984 Olympics was one-sided as, after Tehran (Iran) had dropped out, there remained only one candidate city. It was Los Angeles that submitted a serious bid, which was considered at the 80th **International Olympic Committee (IOC)** meeting in 1978 held in Athens. One reason for the lack of applicants might have been the enormous deficit incurred in the course of staging the 1976 **Montréal Games** in Canada. The Los Angeles Olympics Organizing Committee (LAOOC), with smart planning and careful management, made a profit of over US $200m. and, in turn, restored the popularity of hosting the Olympics. The organizers relied on already existing venues, which then were appropriately renovated/restructured. In fact, the Olympic velodrome and the Olympic aquatic arena, funded largely by 7-Eleven and McDonald's, were the only two new venues constructed specifically for the Games. McDonald's was a key corporate player in supporting the Olympics. They also ran a promotion entitled, *When the US Wins, You Win*. This included giving away free food items. However, due to the non-attendance of world-class athletes from boycotting countries, the promotion almost became a financial disaster for McDonald's as the US Olympic team won far more medals than had been anticipated.

Since the 1984 Olympics were held in the **Cold War** era and immediately after the 1980 **Moscow Games**, they were subject to a variety of political undercurrents. For instance, in response to the US-led **boycott** of the 1980 Games due to Afghanistan being invaded by the USSR a year before, most of the countries (14 including the USSR itself) of the Soviet bloc boycotted the 1984 Olympics. These included Cuba and the Democratic Republic of

Germany (East Germany), which wanted to signal solidarity with their communist comrades. Interestingly, Romania refused to stay in line with the USSR by boycotting the 1984 Games. Romanian athletes were cheered when entering the stadium during the opening ceremony and were successful in the Games, placed second in the unofficial gold medal table. One of the results of the boycott was that the absent countries staged their own (counter Olympic) games, the *Friendship Games* or *Druzhba '84*. Libya and Iran were also absent from Los Angeles for political reasons, but these were more to do with bilateral disputes with the USA than with the boycott initiated by the USSR.

President Ronald Reagan officially opened the Games and Rafer Johnson, winner of the decathlon at the 1960 **Rome Games**, was the final (mystery) torch bearer. He used the Olympic torch to light a specially-built Olympic logo, in which the flame would circle around the five **Olympic Rings**.

The Los Angeles Games were just as eventful in terms of sport achievements. Carl Lewis made his first appearance in the Olympics and won four gold medals—in the 100 metres, the 200 metres, the 4x100 metres relay and the long jump. The first gold medal to be awarded at the Games was also the first-ever medal won by an athlete from the People's Republic of China, Xu Haifeng in pistol shooting. It is worthy of note that following the IOC agreement in 1979 in Nagoya, Japan—the Nagoya Resolution—Taiwan (or the Republic of China) and communist China were permitted to take part in the Olympics as Chinese Taipei and the People's Republic of China, respectively, prompting the Republic of China to boycott the Olympics.

The women's marathon, won by Joan Benoit of the USA, was contested at the Olympics for the first time. These Games also witnessed the Zola Budd–Mary Decker fiasco. The South African-born Budd had received privileged treatment by her newly adopted country, i.e. a fast-tracked passport that allowed her to compete for the United Kingdom at the Olympics, from which as a South African she would otherwise have been banned. She came up against American favourite Mary Decker in the 3,000 metres race. After a strong start and the customary early jostling for position, Decker fell, having tripped over Budd. Budd continued to lead for a while, but eventually finished in seventh place amid a chorus of boos from the spectators. Although the International Association of Athletics Federations (IAAF) found that she was not responsible for the collision, Budd was widely criticized and received controversial press coverage. The use of illegal performance-enhancing substances became a recurring issue at the Los Angeles Games and 11 athletes failed drugs tests.

Louganis, Gregory ('Greg') Efthimios (1960–)

Widely acclaimed as the greatest diver of all time, Greg Louganis was born in El Cajon, California, on 29 January 1960. As a 16 year old, he won a silver medal at the **Montréal Games** in 1976. In 1978 he won a diving scholarship to attend the University of Miami where he studied drama, subsequently transferring to the University of California, Irvine, from where he graduated with a Bachelor of Arts degree. Tipped to win two gold medals at the **Moscow Games** in 1980, he was denied the opportunity to confirm this prediction by the US **boycott**. However, he went on to become a double gold medal winner (in the springboard and high board diving events) at both the **Los Angeles Games** in 1984 and the **Seoul Games** in 1988. During the preliminary rounds in Seoul, he hit his head on the springboard causing bleeding and concussion. In 1987 he was named the US Olympic Committee's Sportsman of the Year, and after the 1988 Olympics he was presented with the Maxwell House/US Olympic Committee Spirit Award for best exhibiting the ideals of the Olympic spirit, demonstrating extraordinary courage and contributing significantly to his sport. In 1994 Louganis made public his homosexuality and the following year published an autobiography, *Breaking the Surface* (co-written with Eric Marcus), in which he claimed to have been involved in a long-term abusive relationship with another man which had led to him becoming HIV positive. Questions were asked within the US Olympic Committee about whether he should have disclosed this information whilst still competing and, with the exception of Speedo, his corporate sponsors deserted him. Louganis was a diving announcer at the 1994 Gay Games and also gave a diving exhibition before a capacity audience. In 1995, in Boston, he was presented with the Visibility Award by US Representative, Gerry E. Studds of Massachusetts. The award honours homosexual men for leadership achievements and, in the following year, he campaigned against the volleyball preliminaries for the Atlanta Games being held in a county in the US state of Georgia that had passed a resolution condemning homosexuals. *Breaking the Surface* was turned into a TV movie in 1997. Having himself experienced problems with drug dependency, since his retirement Louganis has spoken out on behalf of drug and alcohol rehabilitation groups as well as youth clubs and organizations dealing with dyslexia. He has been an actor and has also devoted a considerable amount of time to training dogs for agility competitions. One of the truly great Olympians, it is ironic that Greg Louganis is nowadays probably best remembered for a single misjudged dive and for his sexual orientation.

M

Medal tables

Since the 1908 Games in London when athletes first contested the Olympics as members of national teams rather than as individuals, considerable, and arguably growing, attention has been paid to so-called medal tables. These indicate how many medals (and what type) have been won by competing national teams. Even in this form, the tables can create controversy. Should a team's performance be judged on the basis of the number of gold medals won or on the overall medal total? During the **Cold War**, huge emphasis was placed on the significance of medal tables, particularly by the USSR and the German Democratic Republic (East Germany), but also by other members of the Soviet bloc and, perhaps to a lesser extent, by the USA. Since the end of the Cold War, it has become increasingly apparent that, despite some exceptions to the rule, the medal tables for the Summer Games have been increasingly dominated at the upper end by rich, highly developed (or rapidly developing) countries. Thus, the top five medal-winning countries at the 2008 Beijing Games were the People's Republic of China, the USA, Russia, the United Kingdom and Germany. This type of predictable outcome has prompted the emergence of alternative tables using population size and/or gross domestic product (GDP) as additional criteria. In terms of population size, the 'true' top five countries in Beijing were the Bahamas, Jamaica, Iceland, Slovenia and Norway. Taking GDP as the main criterion, the leading five were the Democratic People's Republic of Korea (North Korea), Zimbabwe, Mongolia, Jamaica and Georgia. Whilst there is a fun element to pitting one table against another, it should be recognized that many governments develop their Olympic funding strategy with medal tables in mind. Targets are set for teams as a whole and for specific sports. Targets are then compared with actual results to determine subsequent funding. Host nations are under particular pressure to perform above their normal standard and regularly do so, although only in certain cases (the Republic of Korea—South Korea—since 1988, for example) is the new standard maintained to a relatively significant degree.

Media coverage

The Olympic Games are one of, if not *the*, most visible and profitable **mega-events** of the modern era. The media, especially television, have played a fundamental role in creating this status and lifting the Olympics on to such a glamorous, global stage. The relationship between the Olympics and the media is long established and symbiotic. The first Olympics to be televised, albeit via closed circuit television and catering only for a local audience, were the **Berlin Games**. The **London Games (1948)** were also televised (by the BBC, which also provided live events) but only to a London-based audience.

The presence and influence of the media have been increasing ever since to the extent that nowadays there are more media personnel at the Olympics than athletes. This process began to most visibly unfold at the 1976 **Montréal Games**, at which the number of journalists, photographers and technicians (approximately 10,000) exceeded the number of athletes (6,028) taking part in the event. Since then the presence of the media at the Olympics has been continuously growing. According to the **International Olympic Committee (IOC)** website, the 2004 Athens Olympic Broadcasting Organization televised more than 4,000 hours of live coverage, utilized more than 1,000 cameras and 450 video tape machines, employed 3,700 personnel and worked with more than 12,000 accredited rights-holding broadcast personnel. This is an impressive scale of preparation, especially given that only 10,625 actual athletes were present.

It is not only the number of broadcasters that has been increasing, but the fees TV channels pay for the rights to broadcast the Games. For instance, the broadcast rights for the Summer Olympics were US $88m. in 1980, a figure that grew to an astonishing $1,700m. in 2008 (the figure is only an IOC estimate and some newspapers go as far as $2,500m.). IOC figures indicate a similar growth in relation to the **Winter Games**. The Lake Placid Games cost $21m. for the broadcasters in 1980, whereas the broadcast rights for the Torino Winter Olympics were sold for an amazing $833m.

Coverage of the Olympics in individual IOC member countries is clearly affected by **nationalism** or, at least, by national interest. Sports in which a specific country is likely to enjoy success are covered to a much greater extent than other sports. Indeed, it has been suggested that in countries such as the USA, media coverage is almost entirely taken up with the exploits of those countries' own athletes.

Mega-events

Mega-events (international parades and global shows/exhibitions) can be defined as recurring, but unique, both globally and locally significant socio-cultural enterprises. They can be major forces in tourism-related development and urban regenerations strategies. Consequently, these events have the ability to considerably affect the economic, political and social landscape of **host cities** and regions and of the events they embrace.

The Olympic Games, one of the most visible contemporary mega-events, have certainly served such a role and have also contributed to a range of changes to the socio-cultural dimensions of sport culture. According to Maurice Roche (*Mega-events and Modernity: Olympics and Expos in the Growth of Global Culture*. London: Routledge, 2000), mega-events are social processes that have idiosyncratic socio-cultural dimensions such as modern/ non-modern, national/ non-national and local/ non-local aspects.

The modern features of the Olympics involve the representation of secular, sometimes techno-rationalist values such as modern technological inventions that have been developed through a process of sport scientization. Non-modern dimensions of mega-events, on the other hand, include the (re)presentation of traditions and the persistence of long-established ceremonies. The Olympics themselves are embedded in antiquity and have a rich culture of historically significant rituals and traditions. For example, the opening ceremonies usually tell the audience stories about the traditions and the past of the host nations. It should also be noted, however, that much of what is presented as traditional, both in relation to the nation and to the Olympics themselves, involves a substantial amount of reinvention and artifice.

The national dimension of the Olympics involves the global presentation of national pride and self-confidence that the host nation experiences when staging the Games. Consequently, mega-events in general and the Olympics, in particular, have the potential to globally disseminate old and new hegemonic messages, thereby often strengthening national pride and nationalistic sentiments. The national pride that many feel when a mega-event is held in their country or when an athlete of her/his nation achieves victory is often considerable. However, mega-events, like the Olympics, are also supranational, involving a cosmopolitan and global outlook and the showcasing and appreciation of multiculturalism. This aspect of the Games, for instance, is demonstrated in the closing ceremonies when the athletes enter the Olympic arena *en masse* and without any strict order, a 'tradition' started at the 1956 **Melbourne Games**.

The local dimension of the Olympics is that they are usually associated with a particular urban space, a city where most of the actual events are staged. Roche observes that 'mega events are localized also in terms of being analysable as urban events, having important and distinctive "urban" level characteristics'. The host city itself often undergoes a total or partial physical transformation that is strategically relevant for the event being held there. This sort of transformation, sometimes taking the form of urban regeneration, has been part of most preparatory work for the Olympics, leading to a range of controversies. For instance, during the preparation of venues for the **Beijing Games,** people were evicted from their homes in order for the sporting arenas to be built. The non-local element of mega-events is that, with the massive assistance of the media, it places a host city and, in turn, its culture, in the global limelight and has the potential to re-position the city in the global inter-city rank order. In sum, mega-events are a feature of modern societies, have multiple dimensions and assume numerous socio-political functions.

Melbourne Games (1956)

Melbourne, Australia, won the right to host the 1956 Olympics by one vote over Buenos Aires, Argentina, also beating Mexico City and six American cities in the process. The Melbourne Games were the first to be held in the southern hemisphere and the location was a major concern, since the reversal of seasons meant that the Games were held during the northern winter. As a consequence, many **International Olympic Committee (IOC)** members were sceptical about Melbourne as an appropriate venue. These worries appeared to be more justified when it became clear that Australian quarantine laws were too severe to allow the entry of foreign horses, thus interfering with arrangements for the equestrian competitions. Those events had to be held separately in Stockholm in June, 1956.

The problems of the Melbourne Games were further compounded by Australian politicians' interference involving petty arguments over financing. For instance, the Premier of Victoria (a state located in the south-eastern part of Australia) refused to allocate money for the Olympic village, which was eventually built in Heidelberg West, and the country's Prime Minister banned the use of federal funds for the Games. By virtue of these initial issues with the Melbourne Games, Avery **Brundage**, IOC President, suggested that Rome, the city selected to host the 1960 Games, was so far ahead of Melbourne in terms of preparations that it might be available as a replacement site by 1956.

Despite initial organizational difficulties, however, the Melbourne Games went relatively smoothly and 3,314 athletes (376 women and 2,938 men) representing 72 National Olympic Committees took part in 145 events. Many athletes performed remarkable sporting feats, but a few stood out and deserve special mention. László Papp of Hungary became the first boxer to win three gold medals. In gymnastics, two athletes excelled: Ukrainian Viktor Chukarin earned five medals, including three gold, to bring his career total to 11 medals, seven of them gold, and Ágnes Keleti of Hungary brought her career total to 10 medals by winning four gold and two silver medals. The US basketball team, led by Bill Russell and K.C. Jones, put on the most dominant performance in Olympic history, scoring more than twice as many points as their opponents and winning each of their games by at least 30 points. There was controversy in the 3,000 metres steeplechase, followed by a selfless display of fair play by the individuals involved. Chris Brasher of the United Kingdom finished first, but judges announced that he was disqualified for interfering with Norwegian Ernst Larsen and they named Sándor Rozsnyói (Hungary) the winner. Brasher appealed against this decision and was supported by both Larsen and Rozsnyói, leading to the reversal of the verdict.

Although the Melbourne Olympics are often referred to as the 'Friendly Games', the degree to which politics interfered with them demands that we re-assess this description. The Games were affected by a **boycott**. Egypt, Iraq and Lebanon refused to participate in response to the Suez Crisis during which Egypt was invaded by Israel, the United Kingdom and France. In 1956 the USSR had brutally crushed the Hungarian Uprising and the Soviet presence at the Games provoked the withdrawal of the Netherlands, Spain and Switzerland. Less than two weeks before the opening ceremony, the People's Republic of China chose to boycott the event because the Republic of China (Taiwan) had been allowed to compete under the name of *Formosa*.

Politics also influenced some of the events and became inseparable from the Melbourne Games. For example, controversial judging prevented the USA from winning all four diving events, which had become almost a tradition. Because Gary Tobian was given suspiciously low scores by the USSR and Hungarian judges, he could only finish second in the platform event. Another memorable, yet controversial and violent, event was what has arguably become the most famous match in water polo history. The so-called 'Blood Bath of Melbourne' was the semi-final match between Hungary and the USSR. The outbreak of violence in the pool was undeniably triggered by the political happenings in Hungary. As the athletes left for the Games, the Hungarian Revolution started, only to be crushed soon afterwards by the USSR's Red Army. Consequently, many of the Hungarian athletes vowed never to return home and felt their only means of fighting back was in the sporting arena.

From the outset, kicks and punches were exchanged during the infamous contest. Towards the end of the match, Valentin Prokopov of the USSR punched Hungary's Ervin Zádor in the face, causing his eye to open up in a bloody gash. The Hungarians were leading 4–0 when the game was brought to an end in the final minute to prevent angry Hungarians in the crowd from reacting to the actions of the Soviet team. Pictures of Zádor's injury were published around the world, with the caption 'Blood in the Water'. Despite the widely held belief that the water actually turned red, this was an exaggeration, albeit one that captured the intensity of the moment. The Hungarian water polo team went on to win the championship by defeating Yugoslavia 2–1 in the final. Half of the Hungarian Olympic delegation defected after the Games as a consequence of the ongoing political events in their home country. In addition, when Olympic officials raised the communist Hungarian flag, numerous complaints were made because it was not the Kossuth Arms flag which had been adopted during the Hungarian uprising. The communist Hungarian flag in the Olympic village was vandalized one night and the communist emblem was removed from the centre and replaced by the Kossuth Arms. When the village staff requested clarification from the newly formed government in Budapest, they were informed that the Kossuth Arms flag was being flown in the city and was, therefore, the correct flag.

The politically loaded sporting climate of 1956 has captured the imagination of writers and directors. A film has been produced entitled, 'Children of Glory' (2008), which shows the Hungarian October Revolution through the eyes of a water polo player and a girl who is one of the student leaders. A similar story can be read in Tibor Fischer's novel 'Under the Frog' (2002).

Prior to 1956, the athletes in the closing ceremony had marched by nation, as they did in the opening ceremony. In Melbourne, the athletes entered the stadium together during the closing ceremony, as a symbol of global unity. In the circumstances, this unity was more symbolic than real.

Mexico City Games (1968)

The year 1968 was momentous for world politics. In May and June of that year, students and workers took to the streets of France and came close to overthrowing the government. Both Robert Kennedy, brother of the late President John F. Kennedy, and Dr Martin Luther King, Jr, one of the leading figures in the American Civil Rights movement, were assassinated. The US Democratic Party's convention in Chicago ended in bloodshed as students and other radicals protested against the escalation of the war in Viet Nam. The USSR's armed forces invaded Czechoslovakia in an attempt to bring to an end reforms associated with what came to be known as the 'Prague Spring'.

The Games of the XIX Olympiad to be held in Mexico City were also affected by the radical spirit of the age. In the lead up to the Games, which Mexico City had won the right to host against opposition from Detroit, Buenos Aires and Lyon, 10 days of student protests culminated in the killing of over 300 protesters by the army and police in the so-called 'Tlatelolco Massacre'. More political controversy was to occur during the Games themselves. During the medal ceremony for the men's 200 metres, two American medallists, Tommie Smith (gold) and John Carlos (bronze) raised black-gloved fists during the playing of the American national anthem. Accompanying them on the winners' podium during their 'Black Power' salute was an Australian athlete, Peter Norman, who demonstrated his support for Smith and Carlos by wearing the Olympic Project for Human Rights badge. Smith and Carlos received lifetime **International Olympic Committee** bans for using the Games to make a political point. Norman was omitted from the Australian Olympic team in 1972. A film, *Salute*, produced by Norman's nephew, Matt, recalls the event and, although Smith and Carlos have latterly fallen out with each other over their respective interpretations of their Olympic race and the protest, in 2008 the two men were jointly awarded the Arthur Ashe Courage Award at the ESPY sports awards, created and broadcast by the American cable television company, ESPN. Another athlete also used medal ceremonies at the 1968 Games to make a political statement, although Vèra Čáslavská's silent protest is much less well known. During the ceremonies for the gymnastics floor exercise and the beam, the Czech woman stood with head bowed and turned away from the flags as the USSR's national anthem was played. One of the greatest ever gymnasts, she was subsequently banned from taking part in sporting events, forced into retirement and denied the right to travel abroad for a number of years.

The venue for the 1968 Games was itself controversial. The first Latin American city to host the Olympics, Mexico City is 7,349 feet (approximately 2,240 metres) above sea level and there were widespread fears about the effects that altitude might have on participants. In the event, the conditions may well have favoured the athletes, not least Bob Beamon, the American long jumper, who jumped 8.90 metres to take the gold medal, setting a world record that would stand until 1991 and an Olympic record that has not yet been surpassed. Africans, some of whom benefited from having lived at altitude, won all the races in men's athletics from 800 metres upwards. Jacques **Rogge**, a future IOC President, competed in the first of his three Olympic Games appearances. During the opening ceremony, Norma Enriqueta Basilio de Sotelo became the first woman to light the Olympic flame. The closing ceremony was transmitted world-wide in colour for the first time.

Migration

When living in the era of migration, the movement of ethnoscapes—the growing significance of people crossing geographic boundaries—has an impact on all aspects of social, cultural and political life. International and global sports and sports events, including the Olympic games, are no exception.

Sports and migrations are intertwined in numerous ways. However, when considering the relationship between the Olympics and migration, four dimensions must be mentioned. These are: i) foreign citizens, or settlers, a type of long-term migrant who gain citizenship and eligibility to represent their host country; ii) employing foreign coaches to train members of the national team; iii) calling upon migrant athletes to return to their home nation; and iv) using the opportunity of competing at an international sporting event to defect and escape an oppressive political regime.

An example of a settler taking part in the Olympics would be Ben **Johnson**, who was born in Falmouth, Jamaica, on 30 December 1961 and emigrated to Canada (Scarborough, Ontario) in 1976. He gained Canadian citizenship and represented his host country at multiple international athletics competitions, including the Olympics. However, because of the drug scandal of which he was at the centre, his elevated standing drastically changed in a short period of time. Within one day, his international and national popularity dramatically declined and his status shifted from hero to villain. Whilst breaking records and winning medals, Johnson, despite his Jamaican origins, was considered a true Canadian hero, a title he lost in a split second after his wrongdoings had come to light.

Nowadays, it is common practice to hire foreign coaches and their experiences and expertise to train club or even national teams. The roots of this practice, however, with regard to the Olympics can be traced back to the 1948 London Games when the British gymnastics team was coached by a German prisoner of war (POW). This person was Helmut Bantz, a Luftwaffe pilot who had been shot down and was still retained as a POW to work on a farm near Leicester. Despite Bantz's support, the British team finished 12th out of 16.

In 1936 the Third Reich called on a few of their athletes who had left Germany because of political developments to return home. A case in point was the high jumper, Gretel Bergmann who, whilst in the United Kingdom, was ordered back to Germany to try for the Olympics team. The Nazi Government wanted her to return in order to help portray Germany as a tolerant country. However, members of her family, who had stayed behind, had been threatened with reprisals if she did not return. She complied, returned to

Germany and competed in the Olympics trials where she outclassed her opponents. Regardless of her performance, though, two weeks before the opening of the **Berlin Games**, she received a letter from the German sport authorities that she would be withdrawn from the German team because her performance was not appropriate for international level.

Zoltán Varga, a Hungarian footballer and member of the national squad that had won the 1964 Olympics in Tokyo, became deeply disillusioned with the direction of Hungarian football and decided to defect, using the 1968 **Mexico City Games** as the opportunity to do so. Since international travel was a privilege and to transfer to a foreign club was unheard of at that time in Hungary, having made the decision to defect, it was vital that Varga should wait for the right moment, which presented itself in 1968. After leaving Hungary, he had a successful career abroad both as a footballer and a coach.

Montréal Games (1976)

The Games of the XXI Olympiad were held in 1976 in Montréal, Québec, Canada, which had successfully competed with Los Angeles and Moscow for the right to become a host city. The Games were seriously affected by political considerations and subsequently by economic difficulties. In terms of politics, over 20 African states staged a **boycott** of the Games in protest against the participation of a team from New Zealand, a country whose national rugby union team had recently toured South Africa, thereby offering legitimacy, it was believed, to the latter's apartheid system. In addition, neither the People's Republic of China nor Taiwan competed as a consequence of ongoing difficulties concerning recognition. After the terrorist attack on the Olympic village at the previous **Munich Games**, there were also concerns about security. These fears were intensified in light of Canada's own domestic political problems with the separatist Parti Québecois (PQ) having made considerable constitutional advances and the Front de Libération du Québec (FLQ) posing a terrorist threat of its own. In 1970 the Front had been responsible for what became known as the 'October Crisis' during which the clandestine, Marxist organization had kidnapped the British Trade Commissioner, James Cross, and killed the Minister of Labour, Pierre Laport. By the mid-1970s, the PQ had become the official opposition party in the Québec legislature. Although the Games passed off without incident, it has been suggested that they were played out within the context of 'culture wars' centred on the issues of nationality and of the use of public expenditure on sport. Were these Canada's Olympics or solely Québec's? The fact that the official opening was conducted by Her Majesty Queen Elizabeth II as Canada's Head of State suggested the former, but of the sports contested only yachting (in Kingston, Ontario) and preliminary football (soccer) matches (in Ottawa and Toronto) took place outside of the province of Québec. As regards public expenditure, the Montréal Games proved to be disastrous, with debts incurred not being finally paid off until the end of 2006. In addition, the Olympic Stadium currently has no major tenant. In sporting terms, the 1976 Games may well best be remembered for the manner in which the rivalry between the USSR and the other Eastern European communist states came to be embodied in the women's gymnastics competition involving Romania's 14-year-old Nadia **Comăneci** and the USSR's Olga **Korbut**, during which the former upstaged her rival by becoming the first gymnast to be awarded the perfect score of 10.0.

Moscow Games (1980)

There were only two bidding cities to be considered by the **International Olympic Committee (IOC)** in 1974. At the 75th IOC meeting, held in Vienna, Moscow and Los Angeles went head-to-head and Moscow, the first Eastern European city to host the Olympics, was given the right to stage the 1980 Summer Games after the first round of votes. Since the **Cold War** was still raging, the Moscow Olympics were subject to a variety of political undercurrents. These Games are famous for the most extensive **boycott** in the history of the modern Olympics. Although the **Montréal Games** had also been affected by a boycott, involving over 20 African states protesting against the participation of a team from New Zealand, the Moscow Games were disrupted by an even larger boycott. This boycott was initiated by US President Jimmy Carter and was a response to the USSR's invasion of Afghanistan in December 1979. In early 1980 President Carter asked the IOC to move the Games to another location. Although the US Congress and the US Olympic Committee (USOC) supported the proposal, the IOC rejected the request. Consequently, the USOC voted to boycott the Games. President Cater issued an ultimatum stating that the USA would boycott the Moscow Olympics if Soviet troops had not withdrawn from Afghanistan by 20 February. The USSR made no attempt to remove its troops and the official announcement confirming the boycott was made on 21 March.

President Carter used his international influence to gain support from other nations for his actions. Some governments, such as those of the United Kingdom and Australia, supported the boycott, but their National Olympic Committees chose to allow the athletes to decide for themselves whether to compete in Moscow or not. However, the USA adopted a different approach, offering no freedom of choice. President Carter went as far as threatening to revoke the passport of any athlete who tried to travel to Moscow. In the end, 67 nations failed to participate in the Olympics, approximately 50 of which were absent because of the US-led boycott. If athletes from absentee countries won medals, their individual achievements were greeted with the Olympic hymn and flag, rather than their own national anthem and flag. Chinese Taipei also boycotted the 1980 Olympics, but was not part of the US-led boycott, being absent because of the 1979 Nagoya Resolution, according to which the People's Republic of China agreed to participate in the IOC if Taiwan (the Republic of China) would be referred to as 'Chinese Taipei'. Instead of taking part in the Moscow Games, many of the boycotting nations participated in the 'Olympic Boycott Games' or the 'Liberty Bell Classic', which were held in 1980 in Philadelphia.

The 1980 Olympics were a tremendous showcase for the Soviet athletes who, aided by the contribution of their partisan crowd which booed all foreign competitors, became a sporting powerhouse, winning an exceptional 195 medals, 80 of which were gold. The German Democratic Republic (East Germany) came second with 126 medals, and Bulgaria was third with 41. Despite the absence of many countries, 36 world records and 79 Olympic records were set during the competition.

One of the highlights of the Games was the exceptional gymnastic performance of Alexsandr Dityatin of the USSR, who won eight medals, including three gold. He was the first athlete to achieve such a feat. Another outstanding Soviet athlete was swimmer, Vladimir Salnikov, who won three gold medals, in the 400 metres freestyle, 4×200 metres relay and 1,500 metres. Cuban super-heavyweight, Teofilio Stevenson, became the first boxer to win the same division three times. On the athletics track, Steve Ovett and Sebastian Coe of the United Kingdom competed against one another in the 800 metres and 1,500 metres in a spectacular rivalry. Ovett won the gold in the 800 metres but Coe, who later became a key figure in London's successful bid to host the 2012 Summer Olympics, took victory in the 1,500 metres.

Another interesting fact about these Games is that over 9,000 drug tests were carried out and none was found positive. This can be interpreted as an indicator of the political climate within which the Moscow Olympics were organized, as we now know that many athletes from the Eastern bloc had taken some form of testosterone to enhance performance.

Owing to the political tensions with which the Moscow Games were associated, the closing ceremony was somewhat untraditional. The handover ritual was altered with the Los Angeles city flag, rather than the US flag, being hoisted to symbolize the next host of the Olympic Games and the Olympic flag being handed over to the IOC President rather than to the mayor of Los Angeles.

Munich Games (1972)

The Games of the XX Olympiad took place in 1972 in Munich, Germany, which had secured the right to host the Games in the face of competition from Detroit, Madrid and Montréal. The first to take place in Germany since the so-called 'Nazi Olympics' of 1936, they were described as the 'happy' or 'carefree' Games and, with the first officially named Olympic mascot, a dachshund called 'Waldi', the aim was to show how far Germany had progressed since the Hitler era. As things turned out, however, the political significance of the Games was the consequence of very different considerations. During the second week of the Games, members of **Black September**, a militant Palestinian organization founded in 1970, murdered 11 Israeli athletes and officials and one German police officer. Holding nine of the Israeli team as hostages, the group demanded the release of over 200 political prisoners detained in Israel together with two German prisoners, Andreas Baader and Ulrike Meinhof, the founders of the terrorist Red Army Faction. The Games were suspended for a day and there was a memorial service at which, it was widely felt, IOC President Avery **Brundage** devoted more of his speech to the **Olympic Movement** than to those who had died. There was also subsequent criticism of the security arrangements in the Olympic village and of the German authorities' rescue efforts. The events themselves were subsequently chronicled in a documentary film (*One Day in September*), and later in Steven Spielberg's 2005 film, *Munich*.

American swimmer Mark Spitz won seven gold medals amidst fears for his safety as a Jew. Two African Americans, Vincent Matthews and Wayne Collett (gold and silver medallists, respectively, in the 400 metres), were banned from future Olympics (the same punishment as had been meted out to Tommie Smith and John Carlos after the 1968 Games) for showing a lack of respect during their medal ceremony. The USSR ended US domination of men's Olympic basketball. In 2008, during the **Beijing Games**, Rabbi Shimon Freundlich led the *Kel Maleh* prayer at a memorial service for the Israelis murdered in Munich.

Muscle gap

The muscle gap—as Jeffrey Montez de Oca calls it—grew directly out of **Cold War** anxieties, especially on the American side, and fundamentally challenged white Western cultural hegemony. It was this challenge that triggered US politicians' apprehensive attitude towards the growing physical superiority of Soviet athletes in the Cold War era. As a result, a general concern emerged in the USA regarding the feminizing effects of modern culture on male citizens. This concern with the decline of traditional masculinity due to modern culture and in the face of the physical dominance of Soviet athletes, coupled with Cold War paranoia, led to perceptions of soft white American male bodies as susceptible to communist occupation. The muscle gap became a real concern in the USA and cultural policy was created to close it. Sport became more than simply an aspect of culture but a set of cultural strategies and tactics to revitalize traditional masculinity amongst male citizens.

Western countries did not exclusively use cultural policies to close the muscle gap but, like some of the countries of the Eastern bloc, increasingly ventured on to the terrain of **illegal performance enhancement**. Although the dominant Western narrative is still that doping in the West is individualized and the unfortunate result of greed, whereas in the East it was a part of a centrally driven political enterprise, in the 1950s drugs were produced and disseminated with the full knowledge of the US sporting authorities. To use whatever technology was available to enhance an athlete's performance was regarded as morally justifiable at that time. Later, Western countries were keen to present themselves as having moved away from the ideologically driven use of banned substances and reprimanded the USSR and the German Democratic Republic (East Germany) for not doing so. Nevertheless, whilst the West was working hard at constructing a moral divide between its athletes and the use of drugs, Western athletes continued to be equally active in the substance-abuse scene.

N

Nationalism

Although commonly described as a global **mega-event**, the Olympic Games remain centred upon international competition. Contestants can only take part as members of national teams, the **flags and anthems** of which are a prominent feature of the Olympic experience.

At the most basic level of analysis, it is easy to see the extent to which sport, arguably more than any other form of social activity in the modern world, facilitates flag waving and the playing of national anthems, both formally at moments such as medal ceremonies and informally through the activities of fans. Indeed there are many political nationalists who fear that by acting as such a visible medium for overt displays of national sentiment, sport can actually blunt the edge of serious political debate. No matter how one views the grotesque caricatures of national modes of behaviour and dress that so often provide the colourful backdrop to major sporting events, one certainly cannot escape the fact that nationalism, in some form or another, and sport are closely linked. It is important to appreciate, however, that the precise nature of their relationship varies dramatically from one political setting to another and that, as a consequence, it is vital that we are alert to a range of different conceptual issues.

For example, like the United Nations, sport's global governing bodies, such as the **International Olympic Committee (IOC)** or the Fédération Internationale de Football Association (FIFA), consist almost exclusively of representatives not of nations but rather of sovereign nation-states. It is also worth noting that pioneering figures in the organization of international sport, such as Pierre de **Coubertin**, who established the modern Olympics in 1896, commonly revealed a commitment to both internationalism and the interests of their own nation-states. Thus, whilst de Coubertin could write enthusiastically about a sporting event that would bring together young (male) athletes from across the globe, he was also specifically concerned with the physical well-being of young French men in the wake of a demoralizing defeat in the Franco–German War.

Much of the literature on the relationship between sport and politics has been concerned with the ways in which nation-states seek to promote themselves, or simply carry out their business, using sport as a useful and highly visible medium. During the **Cold War**, for example, it was apparent that the

USSR and most, if not all, of its Eastern European neighbours used sport in general and especially the Olympic Games to advertise their particular brand of communism. In addition, international rivalry was not only acted out on the athletics track or on the high beam but also impacted on the wider context of events such as the Olympics with the USA seeking to lead a boycott of the **Moscow Games** in 1980 and the USSR and its allies responding in kind when the Olympics moved to Los Angeles in 1984. Related to this is the fact that nation-states also put considerable efforts into acquiring the right to host major events, which are then turned into spectacular exercises in self-promotion by the successful bidders. There can be little doubt that most national leaders in the modern world are highly conscious of the role that sport, and specifically the Olympics, can play in boosting confidence and gaining markers of esteem.

According to writer George Orwell, international sporting competition can best be described as war minus the shooting. The statement is sufficiently ambiguous as to be open to two radically different interpretations. On the one hand, Orwell could be understood to be arguing that international sporting competition acts as a safety valve that makes warfare increasingly less likely. Alternatively, he may have meant that international sporting competition actually keeps alive those very tensions out of which violent conflict is often the inevitable consequence. In fact, we know from other observations that Orwell expressed about sport, that it is the latter reading of his comment which gets closest to an accurate understanding of his meaning. Sport is necessarily competitive and, by implication, conflictual. It is also an important element in the construction and reproduction of social identities. It brings people together. About that there can be no question. It does so, however, in contexts which are arguably more likely to exacerbate tensions than to help to resolve them. This 'fact' of sporting life can be particularly problematic for newly established nation-states, the rulers of which may be inclined to look to sport in their endeavours to foster a sense of national unification.

This has been a common practice in many sub-Saharan African nation-states. It is often the case that alongside national flags and anthems, sporting heroes are of vital importance in helping to promote unity between people who have been brought together within the same constitutional entity that owes its existence far more to the map makers of various European empires than to any collective sense of a shared history. However, using national sporting representatives to this political end can be a difficult strategy to manage in situations in which people retain deep affinities for their own tribal, ethnic or linguistic groups. Perhaps nowhere is this more apparent, and certainly more discussed, than in the 'new' Republic of South Africa where sport has frequently been saluted as the actual, or at least the potential, repository of the collective identity of the 'Rainbow Nation'. It is true that at the symbolic level, few gestures have had more impact than Nelson Mandela donning the shirt of the Springboks rugby union team, so long regarded as the main sporting medium of Afrikaner nationalism. Such gestures notwithstanding, subsequent events in South Africa have demonstrated how difficult it is to unite divided peoples around the banner of national sport. In cases such as these, being accepted (or, in the case of South Africa, being readmitted) by the **Olympic Movement** is a major stage on the road to international recognition and self-esteem. Winning medals takes the process a stage further.

Whilst in most cases those nation-states that constitute international sporting bodies are coterminous with nations, the fact remains that numerous nations throughout the world, as well as other forms of collective belonging, are stateless and are consequently denied representation in international sporting competition just as they are in the corridors of global political power. When considering the relationship between sports and nationalism, therefore, it is important to think in terms both of nation-states and of nations.

In this respect, it is important to note the ways in which francophones in Québec and Catalan nationalists in Spain sought to use the Olympics at the **Montréal Games** and the **Barcelona Games**, respectively, to promote their own political aspirations and, in so doing, dilute the Canadian and Spanish symbolism of the events. The debate about whether the United Kingdom should compete in the association football competition of the London Games (2012) has also been suffused with nationalistic meaning, although, at the formal level, discussions have centred on the future of the separate representation in world football of England, Northern Ireland, Scotland and Wales in the event of such a demonstration of international harmony within a multinational nation-state.

Another long-standing example of the links between nationalist politics and the Olympics is provided by the rivalry between the People's Republic of China and the Republic of China (Taiwan) to receive IOC recognition. The current arrangement whereby Taiwan competes under the name of Chinese Taipei is clearly unsatisfactory in the eyes of those who now regard Taiwan as a separate nation with a demonstrable right to internationally recognized statehood. On the other hand, there are those in Taiwan who remain less convinced and are willing to accept a compromise with China, thereby endorsing the idea that the island is essentially part of China regardless of the political cleavage that currently exists.

Nurmi, Paavo Johannes (1897–1973)

Known together with Hannes Kolehmainen (gold medal winner at the 1912 Olympic Games in Stockholm) as one of 'The Flying Finns', Paavo Nurmi was born in Turku, Finland, on 13 June 1897. At the 1920 Games in Antwerp, he won gold medals in the 10,000 metres, the individual cross country race and the cross country team event, and silver in the 5,000 metres. Four years later, in Paris, he won five gold medals (in the 1,500 metres, 5,000 metres, 3,000 metres team race and both cross country events, setting world records in the first two of these). In 1928 he won gold in the 10,000 metres and silver in both the 5,000 metres and the 3,000 metres steeplechase. Intending to run at the **Los Angeles Games** in 1932, Nurmi was branded as a professional largely at the instigation of the Swedish Olympic Committee and was prevented from doing so. He was recognized as an innovator in terms of training techniques and competitive strategy. During the 1940s he allowed his name to be used to raise money for his country's war effort. He began to suffer coronary problems during the early 1960s and suffered a myocardial infarction in 1967. In 1968 he used some of his personal wealth to establish the Paavo Nurmi Foundation for research into cardiovascular disease. Nurmi died on 2 October 1973 and was given a state funeral. A statue of Nurmi stands outside Helsinki's Olympic Stadium where he lit the flame prior to the Games of 1952. Other statues can be found in his hometown of Turku and at the **International Olympic Committee** headquarters in Lausanne. In 1987 the Bank of Finland issued a banknote depicting Nurmi on one side and the Olympic Stadium on the other.

O

Olympic Charter

The Olympic Charter is the codification of the fundamental principles, rules and by-laws adopted by the **International Olympic Committee (IOC)**, the supreme ruling body of the **Olympic Movement**. It governs the organization and running of the Olympic Movement and sets the conditions for any person or organization directly associated with the Olympic Movement in any capacity. The Olympic Charter clearly outlines that the organization of the Olympic Games is not the only function of the IOC and that the organization is also responsible for promoting specific socio-cultural values and practices relating to sport, in particular, and to wider social and political issues more generally. The complete version of the Olympic Charter can be found on IOC's official website (www.olympic.org).

Olympic Movement

The Olympic Movement consists of the **International Olympic Committee (IOC)**, National Olympic Committees, International Federations (IFs) and other organizations that are officially recognized by the IOC and agree to be guided by the **Olympic Charter**. The Olympic Movement was established in 1894 by Pierre de **Coubertin** in making his attempt to revive ancient Greek ideals and apply them in modern times. His admiration for antiquity derived from the classical education he had received as a member of the French aristocracy. In his opinion, ancient Greece, its culture and its lifestyle represented the golden age of humankind whereas the France of his time was part of a decadent civilization. In order to reverse this physical and moral decline, he turned his attention to reforming France's physical culture by combining modern sporting practices with ancient beliefs. The linking of ancient and modern provided the foundation for the Olympic Movement, the underpinning philosophy of *la pédagogie sportive* as de Coubertin called it. This is a humanistic endeavour aimed at the pursuit of ethical and international/global harmony. That is, the Olympics were to become a kind of sporting Esperanto (a constructed international auxiliary language), which would uphold and cherish positive moral values channelled through sport to create and perpetuate peace.

According to the IOC's website (the Olympic Charter outlining the Olympic Movement and its actions can be found at www.ioc.org), the Olympic Movement is concerned with four main goals: balanced body-mind development, finding joy in effort, educating good role models, and respect for universal ethics and for others. In short, the Olympic Movement's overall aim is to promote the idea of Olympism through sports thereby attempting, at least at the level of symbolism, to elevate the value of human dignity and build a peaceful society.

In principle, these ideals represent a humanistic attitude to sport, competition and society as a whole. In reality, however, there is a rich tapestry of evidence that indicates self-centred and discriminatory attitudes often being followed and adopted by the IOC. For instance, during the 1972 **Munich Games**, after the **Black September** acts of **terrorism**, the Games were suspended for a day and a memorial service was held for the victims. At this service, it was widely felt that the IOC President, Avery **Brundage**, devoted more of his speech to the Olympic Movement than to those who had died, thus displaying an Olympic-centred attitude. Another example is provided by the IOC's long-standing preservation of male hegemonic dominance in the Olympics. Until the early 1980s, the IOC was an exclusively male preserve. The first two female members were admitted in 1981 but, although the number of women involved in the IOC has gradually increased, the organization remains predominantly male. This scarcely reflects a humanistic approach to social development.

Olympic oath

The Olympic oath is one of the rituals of the modern Games and has its roots in antiquity. The modern Olympic oath was originally written by Pierre de **Coubertin** and was first taken at the 1920 **Antwerp Games** by Belgian fencer, Victor Boin, who recited the following words:

> We swear that we will take part in the Olympic Games in a spirit of chivalry, for the honour of our country and for the glory of sport.

Since then, the Olympic oath has become one of the fundamental rites of the modern Games. One athlete from the host country takes the oath at the opening ceremony on behalf of all the participating athletes, whilst holding a corner of the Olympic flag in their left hand and raising their right hand. The athletic oath has been modified over time to reflect cultural changes and the changing nature of the sporting event. The current version of the oath, addressing the growing issue of doping, was introduced in 1999 and was first taken at the **Sydney Games** in 2000.

> In the name of all competitors I promise that we shall take part in these Olympic Games, respecting and abiding by the rules which govern them,

committing ourselves to a sport without doping and without drugs, in the true spirit of sportsmanship, for the glory of sport and the honour of our team.

It is not only athletes who are required to take the Olympic oath, but judges as well. This ritual was first introduced at the 1972 Munich Games. As with the athletic oath, a judge from the host country recites the Olympic creed on behalf of all the participating judges. Although there have been many permutations, the fundamental message has remained:

In the name of all the judges and officials, I promise that we shall officiate in these Olympic Games with complete impartiality, respecting and abiding by the rules which govern them, in the true spirit of sportsmanship.

Olympic rings

The symbol of the Olympic Games is composed of five interlocking rings, coloured blue, yellow, black, green and red on a white background. The five rings stand for passion, faith, victory, work ethic and sportsmanship. Despite the popular view, according to which the rings have their links in antiquity, there is in fact no connection between the rings of the modern Olympics and the ancient Games. The rings were originally designed in 1912 by Pierre de **Coubertin** in preparation for the 1914 Olympic Congress, which had to be suspended due to the outbreak of the First World War. Nevertheless, the symbol was later adopted and its official debut took place at the **Antwerp Games** in 1920. As regards the colours, it has been suggested that originally de Coubertin intended the rings to represent the first five host nations of the modern Games (Greece, France, the USA, the United Kingdom and Sweden). The colours of the rings (and the white background) include all the colours of the first five host countries' flags. It is also suggested that de Coubertin's original intention was to insert an additional ring for every new nation that subsequently hosted the Olympics. This may well have been de Coubertin's intention. However, what we know for certain is that according to the **Olympic Charter**, the rings now represent the union of the five continents and the meeting of athletes from throughout the world at the Olympic Games. In this sense, the Americas are viewed as a single continent, Antarctica is absent and, hence, the five continents are taken to be Africa, America, Asia, Europe and Oceania.

From a commercial perspective, the Olympic rings are one of the Olympic symbols and are the exclusive property of the **International Olympic Committee (IOC)**. They are protected around the world in the name of the IOC by trademarks or national legislations and cannot be used without the IOC's prior written consent.

Owens, James Cleveland ('Jesse') (1913–80)

Jesse Owens was born on 12 September 1913 in Oakville, Alabama, the seventh child of a sharecropper. Due to the inhumane conditions endured by sharecroppers at that time, in the early 1920s the Owens family moved to Cleveland, Ohio to begin a new life. It was there that Jesse Owens's physical ability was discovered by Charles Riley, who launched his athletic career. By the early 1930s, Jesse had a state-wide reputation based on his athletic performances; international recognition came during the 1936 Berlin Olympics. The **Berlin Games** placed Jesse Owens in the international spotlight partially because of his athletic performance and partially because he was not simply competing against other athletes but in the face of a well-orchestrated racist political campaign, spearheaded by the *Führer* (leader) of Germany, Adolf Hitler. Owens almost single-handedly undermined the Nazi ideology of Aryan physical supremacy by winning four gold medals at the Olympics (the 100 metres, 200 metres, 4x100 metres relay and the long jump), beating many German participants in the process. Although the victories of Owens and other black athletes infuriated Hitler, the widely accepted 'fact' that Hitler snubbed Owens has no truth in reality.

The political significance of Owens's exceptional athleticism, however, does not end with his struggle against German Aryan racial domination. Like most African Americans at the time, Owens was perceived as a second-class citizen in his home country, which, it could be argued, was scarcely less racist than Nazi Germany at the time. While travelling with his white team-mates, Owens was often segregated and had to stay in 'Blacks Only' hotels and eat in 'Blacks Only' restaurants. Even after his outstanding Olympic success, he never received any endorsement deals nor was he invited to the White House. Owens's athletic accomplishments were only officially recognized in 1955 by President Eisenhower who granted him the 'Ambassador of Sports' title and, subsequently, in 1976 (four years before his death) by President Ford who awarded him with the Medal of Freedom. Unfortunately for Owens, these privileges did not solve the financial difficulties with which he had to struggle throughout his life. He died of lung cancer on 31 March 1980 in Tucson, Arizona. The Jesse Owens Memorial Park in Oakville was opened in 1996.

P

Paris Games (1900)

Although Pierre de **Coubertin** wanted Paris to host the first modern Olympics in 1900, the newly founded **International Olympic Committee (IOC)** believed that an earlier date would be more beneficial for the movement. Therefore, not the first but the second modern Olympics were held in Paris in 1900 as part of a world fair or exposition, *Exposition Universelle Internationale*. The exposition organizers spread the events over five months and, thus, underplayed the significance of the Games. The Olympic aspect of the sporting events was presented as being so insignificant that some of the competing athletes did not know that they had actually competed at the Olympics. In other words, these Games suffered the same neglect as the **St Louis Games** would four years later and were reduced to a mere sideshow of the exposition, lost in the chaos of other cultural activities.

At these Paris Games, no opening and closing ceremonies were held and most of the winners did not receive medals; cups or trophies were presented instead. These were the first Olympics at which women were allowed to compete and Charlotte Reinagle Cooper, a British tennis player, was the first woman to become an Olympic champion. The Olympic programme consisted of 95 events, in which 997 athletes (22 women and 975 men) from 24 National Olympic Committees took part.

Though these were the second modern Games, they were no better organized than the first. The conditions athletes had to face were appalling. For instance, scheduling problems were so common that many contestants never made it to their events. When they did arrive on time, they often found conditions unfit for an international sporting event. For example, the area for the running events was covered with grass rather than cinder and was dangerously uneven. The competitors in the throwing events did not even have enough room to throw. Furthermore, the hurdles were made out of broken telephone poles and the swimming events were conducted in the River Seine with its extremely strong current. The problems did not end there. The marathon resulted in controversy as participants began to suspect that the French competitors had cheated, since the American runners reached the finishing line without seeing other athletes pass them, only to find the French runners already at the finishing line and seemingly refreshed. Equestrianism made its Olympics debut but then disappeared until 1912.

Due to the marginalized role of the Paris Olympic Games and the devaluation of the Olympic ethos, the IOC proposed the idea of holding interim or
intercalated Games in Athens in 1906. Interestingly, although the **Athens
Games (1906)** may have helped to keep the entire movement alive, today the
IOC does not recognize the interim Athens Olympics as Olympic Games and
does not regard as official any of its events and achievements, such as new
records and medals.

Paris Games (1924)

Despite competition from Los Angeles, Rio de Janeiro, Amsterdam and
Rome, the Games of the VIII Olympiad were awarded to Paris which, thus,
became the first city to officially host the modern Olympics for a second time.
These were the last Olympics to be organized during Pierre de **Coubertin**'s
Presidency and he had undoubtedly used his personal influence to ensure that
the Games returned to France. The Games were attended by 44 nations
although Germany was still excluded in the aftermath of the First World War.
Ireland (as the Irish Free State) took part for the first time as an independent
nation-state. The Olympic motto—'*Citius, Altius, Fortius*' (Faster, Higher,
Stronger)—was used for the first time. This was also the first time that there
was an official Olympic village and the closing ceremony included the raising
of three flags, those of Greece, France as the host, and Holland where the
next Games were scheduled to take place in the city of Amsterdam. As a
consequence of suspicions about professionalism, tennis made its last Olympic
appearance until its return at the 1988 **Seoul Games**.

The 1924 Games witnessed some impressive athletic performances not least
those of Paavo **Nurmi**, the so-called 'Flying Finn', who won five gold medals.
Paris was also the scene of gold medal winning performances by the United
Kingdom's Harold Abrahams in the 100 metres and Eric Liddell in the 400
metres, their exploits being later dramatized in the 1981 film, *Chariots of Fire*.
Another gold medallist was Johny Weissmuller who subsequently achieved
even greater fame for his screen portrayal of Tarzan. William deHart Hub-
bard became the first black athlete to win a gold medal in an individual event,
the long jump.

Prolympism

The term prolympism denotes the relatively recent merger of initially two
separate sport systems (professional and Olympic) into a monoculture. It
refers to the global diffusion, production and consumption of sports. Pro-
lympian sporting practices are often perceived as linked to aggressive, neo-
colonialist, sports culture formations that will eventually lead to the eradica-
tion of alternative and traditional sporting practices, thereby creating global
sporting homogeneity. In other words, this is seen by some as an exclusive
and exploitative globalizing sporting system (see **Globalization**) that has

cultural, ideological and structural features which will eventually allow it to dominate the global sport and Olympic arenas.

Protests

Protests relating to and involving the Olympic Games have taken many forms and have been used for a variety of purposes. Within the context of the Olympics themselves, the **boycott** has been the most prominent form of protest, with **International Olympic Committee (IOC)** member states expressing political grievances by refusing to participate in particular Games (see **Moscow Games** and **Los Angeles Games**). The high profile of the Olympics has also allowed individuals and groups to make political statements. These have ranged from the 'Black Power' demonstration by African American medal winners, Tommie Smith and John Carlos, at the **Mexico City Games**, to the murderous actions of **Black September** during the **Munich Games**. Significant protest movements can also emerge in the lead up to particular Games. Sometimes these focus on the Games themselves and especially on the cost of being a host city and the planning implications of building for the Games. The latter was certainly a concern in relation to the 2008 **Beijing Games** and also the **Sydney Olympics** of 2000, when anti-Olympic protesters also focused on racism and social injustice in Australian society and subsequently on the imposition of restrictions on the right to protest. In Japan, an anti-Olympic movement grew out of protests in Nagoya against that city hosting the 1988 Games which were eventually awarded to Seoul, Republic of Korea (South Korea). However, other major protest movements have been less concerned with the Games themselves, but have nevertheless exploited the attention focused on specific **host cities**. Thus, in 1968 student protesters supporting a variety of social and political reforms took to the streets of Mexico City resulting in the so-called 'Tlatelolco Massacre' in which over 300 hundred students were killed by the police and army. So serious was the crisis that the IOC considered cancelling the Games. Similarly, in 1987 student protesters demanding democratic elections in South Korea put the 1988 **Seoul Games** in jeopardy. Their protests and, therefore, indirectly the significance of the Olympics, helped to bring about the removal from office in December 1987 of President Chun Doo-whan, thereby opening the way for South Korea's transition to democracy. More recently, during 2008, the **torch relay** preceding the Beijing Games became the focus for world-wide protest with demonstrators concerned with China's treatment of Tibet and with **human rights** violations within the country as a whole.

R

'Race' and racism

Whilst there may well have been racial discrimination in relation to team selection for the Olympics in many parts of the world and over many years, the **Olympic Movement** has been most directly associated with and affected by 'race' issues in three main ways.

First, there have been debates within the **International Olympic Committee (IOC)** about institutionalized racism within member countries. In the case of Nazi Germany, a proposed boycott of the **Berlin Games** was avoided despite widespread knowledge of the ruling regime's racist policies which extended into the world of sport itself. Similarly, the **Beijing Games** of 2008 went ahead regardless of concerns about the Chinese Government's treatment of ethnic minorities. In the case of South Africa, on the other hand, after the passing of a United Nations General Assembly Resolution in 1962, the IOC agreed to take part in an international **boycott** in protest against that country's racist apartheid policy. Having competed in the Olympics for the first time in 1904 and having participated in the **Rome Games** two years earlier, South Africa was prevented from taking part in further Olympics until 1992, by which time the apartheid system was in the process of being dismantled.

Second, the Olympic Games have been used as a stage upon which to protest directly against racism. The so-called 'Black Power' salute of Tommie Smith and John Carlos at the **Mexico City Games** was aimed at ongoing racial discrimination in the USA and in support of the campaign for Civil Rights for Americans of colour. Much earlier, the African American hero of the **Berlin Games**, Jesse **Owens**, is said to have claimed that it was not Hitler's response to his achievements that upset him but that of his own President, and Cassius **Clay** (Muhammad Ali) was clearly unhappy about returning to the USA as an Olympic hero only to be treated as a second-class citizen. Kathy **Freeman**'s performance at the **Sydney Games** and her subsequent celebration which involved both an Aborigine flag and an Australian one might be interpreted as a less direct, more subtle, comment on racist practices in her home country.

Finally, discussions within the academic community about Olympic performances have also raised the spectre of 'race' with some scholars arguing the case for a modified form of biological determinism with reference to the success of sprinters of west African origin and distance runners from north and east Africa. Such arguments have been refuted (on occasion even described as racist) by those who fear their wider implications and argue instead that socialization, role models and access are more important explanatory factors.

Revithi, Stamata

Due to the strength of patriarchy, women were not allowed to compete at the first modern Olympics. However, there was an unofficial female competitor in the marathon in the **Athens Games (1896)**. A Greek woman, Stamati Revithi, was refused the right to compete in the men's marathon race and decided to run by herself the next day. Against all odds, she finished the course but the final lap had to be completed outside the stadium as Revithi was refused entry. Since athletics officials could not remember her name they labelled her (and, perhaps, female athleticism in general) *Melpomene*, the Greek muse of Tragedy.

Rogge, Jacques (1942–) (IOC President: 2001–)

Jacques Rogge, Belgian by nationality and an orthopaedic surgeon by profession, was elected to be the eighth **International Olympic Committee (IOC)** president in July 2001, replacing Juan Antonio **Samaranch**. As an athlete, Rogge competed in the yachting competitions at the **Mexico City Games**, **Munich Games** and **Montréal Games**. He was also a member of the Belgian national rugby team. Before being elected as IOC President, he served as president of the Belgian National Olympic Committee from 1989 to 1992. He became president of the European Olympic Committees in 1989, an IOC member in 1991 and an IOC Executive Board member in 1998.

Rogge was elected president in Moscow at the 112th IOC meeting in the hope that he would improve the IOC's image, which had been tarnished by scandals, particularly the resignations and criminal charges of bribery that were associated with the successful bid (see **Bidding**) by Salt Lake City for the 2002 **Winter Games**.

Rogge's IOC presidency faced problems from the outset. In addition to the bribery-related scandals surrounding the IOC and the **Olympic Movement**, Rogge also had to deal with the consequences of the 9 September 2001 terrorist attacks in the USA. Due to the possibility of a further terrorist attack on US soil, Rogge was granted emergency powers to cancel the Salt Lake City Games without the calling for a vote by the full committee. He had no intention of retreating, though. On the contrary, he openly expressed his confidence in the safety of the event and decided that he would stay in the Olympic village, along with the athletes, during the 2002 Olympics. His approach to the Olympics and to sport generally was welcomed by the critics of the Olympic Movement and a brighter, less scandalous future for the IOC was predicted. In particular, his attitude towards doping and to hosting the

Olympics has altered views about the IOC. In terms of staging the Games, Rogge has sought to create more possibilities for developing countries to successfully submit bids. He has argued that developing countries would soon be able to realistically submit bids through government backing and new IOC policies that constrain the size, complexity and cost of hosting the Games. However, given that the Olympics have continued to grow and become more expensive with each quadrennium, Rogge is yet to convince the sceptics about the feasibility of his ideas.

After confronting the possibility of a terrorist attack on the 2002 Winter Games, Rogge faced more problems with regard to the **Beijing Games**, which revolved around **human rights** and media censorship. Although he expressed his concerns about Tibetan unrest prior to the Olympics, Rogge failed to fully acknowledge the possibility of an extensive boycott. Furthermore, he clearly believed that the Olympics would provide a unique and long-awaited opportunity for journalists in the People's Republic of China to publish their views without censorship. However, it soon came to light that the Chinese Government would not relax its tight media control easily. In July, Kevan Gosper, IOC spokesman, announced that the internet would indeed be censored for journalists in Beijing for the Summer Olympics. There was also a reference to a secret deal between the IOC and Chinese officials regarding censorship, but Rogge denied that such a meeting and agreement had taken place. In addition to the internet fiasco, Rogge publicly criticized the gestures of Usain Bolt (the Jamaican sprinter and gold medal winner in the 100 metres, 200 metres and 4×100 metres relay) after the men's 100 metres final and described his actions as unsportsmanlike. Rogge's views generated a media flurry and were perceived as out of touch and myopic.

By virtue of the above, it seems that, despite the initial hopes for his IOC presidency, Rogge now follows a similar path to that of his predecessors, generating the sort of controversy that has been historically so often associated with IOC presidents.

Rome Games (1960)

Having originally been given the right to host the Summer Olympics of 1908, Rome finally played host in 1960—the Games of the XVII Olympiad. Use was made of a mixture of ancient and modern venues for the Games, played out against the wider backdrop of the **Cold War**, which created a general atmosphere of suspicion and ill will. These were the first Games to be fully televised world-wide courtesy of CBS, which had acquired the broadcasting rights. South Africa made what was to be its last Olympics appearance until 1992. Singapore competed for the first time under its own national flag, having been granted the right to self-government by the United Kingdom a year earlier. Nationalist China (Republic of China/Taiwan) was forced to participate, under protest, as Formosa. There were some notable and politically significant triumphs for individual competitors. Cassius **Clay** (later

Muhammad Ali) won gold in the light heavyweight boxing competition and another African American, Wilma Rudolph, a sprinter who had suffered from polio as a child, won three gold medals on the track. The future King Constantine II of Greece won a gold medal in sailing. Perhaps most importantly, at least in terms of the future of Olympic middle- and long-distance running, the marathon was won by Abebe Bikila of Ethiopia (still referred to in the sports media of the time as Abyssinia) who, running barefooted, became the first black African to win an Olympic gold medal and, in so doing, became a role model for subsequent generations of north and east African runners. In terms of the future of the Games more generally, arguably the most worrying incident was the death of Danish cyclist, Knud Enemark Jensen, who collapsed during the road race and fractured his skull. Some commentators blamed the heat and questioned whether Rome in mid-summer was an appropriate venue for the Games. However, it was discovered that the dead cyclist had taken a performance-enhancing drug. Indirectly, this incident led to the **International Olympic Committee** establishing a medical commission in 1967 and introducing drug-testing at the 1968 Winter Olympics in Grenoble and the 1968 Summer Games in Mexico City.

S

Samaranch, Juan Antonio (1920–) (IOC President: 1980–2001)

Like all the **International Olympic Committee (IOC)** presidents before him (and probably for many years to come), Juan Antonio Samaranch was born into a wealthy family in Barcelona where he later studied commerce and business studies. He also studied in the United Kingdom and in the USA.

Samaranch became an IOC member in 1966. Before becoming IOC President, he assumed various positions within the **Olympic Movement** such as head of protocol (1968–75, 1979–80), member of the executive board (1970–78, 1979–2001), IOC Vice-President (1974–78). He was then elected IOC President in 1980. Samaranch assumed his IOC presidency in troubled times, just after the **Moscow Games** and before the **Los Angeles Games**, both of which have become infamously associated with the political tactic of **boycott**. His response was to govern the IOC with an iron grip. During his presidency, he undeniably improved the fortunes of the Olympic Movement and developed a popular and commercially successful product. Nevertheless, although he saved the Olympics from potential bankruptcy, he was often accused of excessive commercialization of the Games.

During Samaranch's period in office, the IOC reduced the amateur rule to the status of a mere formality (see **Amateur code**). Consequently, athletes, such as Paavo **Nurmi**, who had been stripped of their medals and record-breaking achievements were reinstated. Another notable example of the IOC relaxing its policy towards accepting the eligibility of professional athletes was the 1992 US men's basketball team, known as the 'Dream Team', which included numerous National Basketball Association (NBA) players.

Samaranch was frequently criticized for a range of issues during his tenure at the IOC. He was renowned and criticized for his expensive lifestyle, especially his habit of staying in luxury hotels. A few IOC members disagreed with such observations. We may fail to draw an unequivocal conclusion regarding the truth of Samaranch's expensive lifestyle; however, it is known for certain that he enjoys and demands subservience. Whilst IOC President, he insisted that he be addressed as 'Your Excellency'.

Another controversial issue surrounding the presidency of Samaranch centred on his attitude to the use of performance-enhancing substances. His approach to doping was revealed by Dick Pound, former President of the World Anti-Doping Agency and former IOC Vice-President, who said that without the Festina fiasco in cycling, drugs-related issues would have continued without intervention.

Samaranch's political affiliation was also a matter of interest to his critics. Andrew Jennings, the investigative reporter, argued in a documentary and later in a book that Samaranch was a Nazi sympathizer by exposing direct links that he had had to Nazi organizations and to the dictator of Spain, Francisco Franco (these links have been referred to by academics also). Jennings presented a range of evidence in support of his argument and still received a five-day suspended jail sentence (or, as he termed it, 'the Samaranch Prize for Literature') meted out by a Swiss court.

During his presidency, it became a tradition that when giving the presidential address at the close of each Summer Olympics, Samaranch would praise the organizers by stating that they had staged 'the best ever Games'. However, in his closing presidential speech in 1996, he called the **Atlanta Games** 'most exceptional', thereby breaking with his own tradition. This was probably an intentional response to some of the political and commercial tensions between the IOC and the Atlanta Committee for the Olympic Games. Interestingly, four years later Samaranch returned to the familiar pattern and called the 2000 **Sydney Games** 'the best ever'. In 2001 Samaranch was succeeded by Jacques **Rogge**, the current IOC President. After his resignation, he was awarded the title of Honorary President. It was widely believed that Samaranch's efforts on behalf of the Olympic Movement would be rewarded in 2009 with the selection of Madrid as host city for the 2016 Summer Games. In the event, however, although reaching the final stage of voting (having beaten Tokyo and Chicago), the Spanish capital lost out to Rio de Janeiro.

Security

To date, the Olympics have been affected by two different types of security threat. The first of these is the threat posed to the Games by citizens of the host country opposed to the staging of the event for a variety of political and social reasons. Resultant protests have often been dealt with in a ruthless manner by state security agencies. This was particularly apparent in 1968 during the lead-up to the **Mexico City Games** and also prior to the **Seoul Games** in 1988. At other recent Games, although the security presence may have been less lethal, various means have been used to silence dissident voices, most notably for the **Beijing Games** in 2008. The second type of security threat to the Games involves the use of the high profile of the event itself to promote political causes that are unconnected to the Games themselves. The most dramatic example of this was the taking and killing of hostages by Palestinian terrorists during the **Munich Games** in 1972. The cost of securing the Games against actions of this kind has risen dramatically in the wake of the 9 September 2001 terrorist attacks in the USA. Indeed, it is likely that security will prove to be the major concern for the organizing committee of the London Games (to be held in 2012), given that there have been terrorist attacks carried out by Islamic fundamentalists in London in the recent past and in light of the fact that it is almost certain that British forces will still be engaged in military operations in Afghanistan by the time the Games take place (see **Terrorism**).

Seoul Games (1988)

The Games of the XXIV Olympiad took place in Seoul, the capital city of the Republic of Korea (South Korea). With a significant **boycott** having adversely affected each of the three previous Olympics, the **International Olympic Committee (IOC)** was keen that the 1988 Summer Games should attract as many entries as possible. However, the choice of Seoul was not without political controversy for two main reasons. First, the Korean peninsula had remained divided since 1945, a period which had also seen war between the Democratic People's Republic of Korea (North Korea) and South Korea aided and abetted by their respective **Cold War** allies. Not surprisingly, this prompted questions about how not only North Korea but, perhaps more importantly, its communist friends would react to a decision in Seoul's favour. Second, when the decision to choose Seoul as host city was actually taken, South Korea was under authoritarian, military rule. Clearing Seoul's path, however, was the fact that the only rival was Nagoya in Japan. Unlike Seoul, the Japanese city received only limited national support for its bid and was also confronted by significant local opposition. Seoul, on the other hand, was enthusiastically backed by the South Korean Government, with its President clearly recognizing the propagandist potential of South Korea becoming only the second Asian country to host the Games, and by the country's major economic enterprises, the executives of which saw the Games as providing an opportunity for them to establish themselves in the global marketplace. Once Seoul was awarded the Games, attempts were made both by the South Koreans and by the IOC to meet North Korean demands to be involved. In the end, however, this proved impossible and the North Koreans were the most significant of a very small number of absentees when the Games eventually took place. Before that, in an attempt to destabilize South Korea in the lead up to the Games, in 1986 North Korean agents planted a bomb at Kimpo airport, the major international airport near Seoul, killing five people and injuring another 30. The attack damaged South Korea's reputation to a degree, as it happened just before the opening of the Asian Games that were also held in Seoul. However, when North Korean secret agents then attacked a South Korean airplane in 1987, killing all 115 passengers, sympathy was expressed for the forthcoming Olympic hosts and the rest of the world, including the Soviet bloc, began to question North Korea's criticism of South Korea. Thus, while the communist regime succeeded in attacking the South before the Olympics, it failed to persuade the international community to support its position. Indeed, North Korea received little or no support from the USSR or the overwhelming majority of its satellites, which were eager to take part in the Games for sporting and propagandist reasons. In 1987 the motto of the Seoul Games, emphasizing 'Harmony and Progress', proved more prophetic than might initially have been anticipated when internal opposition, using the oxygen of publicity provided by the forthcoming Games, and external pressure combined to ensure the removal of the existing government and the introduction of partial democratization. Thereafter, the Games played a major part in allowing South Korea to become more globally recognized and understood. In addition, those leading companies which had helped to make the Games possible reaped substantial economic benefits. South Korean participants ensured that the host country finished fourth in the overall **medal tables**. This achievement helped to lay the foundations upon which South Korea has gone on to be a consistent Olympic performer throughout subsequent years.

Sex test

Gender verification in sport was the direct result of **Cold War** paranoia and involved 'verifying' the eligibility of an athlete to compete in a sporting event limited to a specific gender. This practice reflected the attitude of Western politicians, in general, and US politicians, in particular, towards the growing physical superiority of Soviet and Eastern European athletes, especially female athletes, during the Cold War era. As evidenced in debates about the so-called **muscle gap**, there was general concern in the USA that modern Western culture feminizes its male citizens and this was compounded by the masculine appearance of successful female athletes of the Eastern bloc. The effeminizing impact of modern culture and the physical dominance of communist female athletes fundamentally shook Western gender stereotypes. It was not uncommon for world-class athletes from communist countries to be ridiculed and called lesbians or men in disguise in the Western press. Even though there was only a handful of high-profile cases, the Cold War-based politics of sex had a significant effect on international sports.

In an attempt to restore the 'natural' gender order and its roles and also Western sporting dominance, a sex test for women taking part in international competitions was introduced and first carried out in 1966 at the European Athletic Championships and in the 1968 **Mexico City Games**. Initially, the test consisted of female athletes undressing before a panel of physicians. This procedure was considered demeaning and was replaced by sex chromatin analysis of cells from a buccal smear to check for double X chromosomes, a chromatic indicator of femininity. Chromatic analysis was replaced with polymerase chain reaction (a DNA enzyme-based testing) by the **International Olympic Committee (IOC)** in 1992. Although this procedure was less humiliating for competitors, experts argued that the test was pointless and had the potential for causing great psychological damage to women who, sometimes unknowingly, have certain disorders of sexual differentiation. Therefore, the

1996 **Atlanta Games** were the last ones at which the sex test was carried out. The IOC officially stopped the sex verification examination in 1999 and the 2000 **Sydney Games** were sex test free. However, the controversial gold medal winning performance by South Africa's Caster Semenya in the women's 800 metres at the 2009 World Athletic Championships in Berlin has once again brought the issue into the forefront of public attention. Whether or not Semenya competes in the London Games in 2012 will surely be a matter for intense debate within the IOC.

Spectacularization

Critics of the Olympic Games have tended to reproduce arguments used in relation to the use of athletic contests in the ancient world. Thus, the term 'bread and circuses' is frequently invoked to imply that the Games, along with other sporting **mega-events**, are used by governments and elites more generally to pacify the watching public. For this to be successful, equally important are high-level performances in the Games themselves, innovative architecture and increasingly spectacular opening and closing ceremonies. The need for records has undeniably been a factor in some performers, with or without the backing of their respective nation-states, breaking the rules of their particular sports. The design and construction of venues have become major cost factors in planning for the Games as have the arrangements for the ceremonial events. In many cases, opening ceremonies in particular are used politically to suggest national unity in societies in which there is considerable ethnic difference and disharmony. This was certainly the case in 2008, although it is also widely accepted that the **Beijing Games** achieved a standard in terms of Olympic architecture and ceremony which no subsequent host city will be able to match. It is unlikely, however, that the pressure to put on the greatest show on earth will recede entirely.

Stadiums

Although Olympic activities take place at numerous venues, the Summer Games invariably have as their focal point a main stadium which hosts the opening and closing ceremonies and also track and field events. Many stadiums were constructed specifically for the Games and have the word 'Olympic' in their title. Almost as many others, however, had already been constructed, some of them having the word 'Olympic' added to their names after the bidding process has ended in success for the host city. Some stadiums that were already built simply retained their original names. Only one stadium has been used twice as the main Olympic arena—the Los Angeles Memorial Coliseum. However, both the Panathinaiko Stadio (Athens) and the Vélodrome de Vincennes (Paris) have hosted events at two separate Summer Games. In addition, the Melbourne Cricket Ground was used for the **Melbourne Games** of 1956 and then hosted a football (soccer) match as part of the **Sydney Games** of 2000.

In some cases, an Olympic stadium can become a vital element in an urban regeneration programme. The experience of the **Seoul Games**, for example, is that the area around the Seoul Olympic Stadium has become much sought after and property prices have escalated. Urban renewal is also a key feature of planning for the London Games in 2012, although, as in the case of other successful and even unsuccessful bids, there is some uncertainty as to what this means in practice for current residents of the area in which the main venues will be constructed.

Arguably the stadium built for the **Beijing Games** of 2008 is the most celebrated to date. Designed by the Swiss architecture firm, Herzog and de Meuron, the Beijing National Stadium (or 'The Bird's Nest' as it is popularly known) is the world's largest steel structure. There are already fears, however, that the stadium will not be used consistently in the future. Indeed, this has long been a major concern about Olympic stadiums and helps us to understand why the stadium to be built for London 2012 will see its capacity reduced from 80,000 to 25,000 after the Games have taken place.

Some Olympic stadiums—such as those in Stockholm and Helsinki—are now much-loved, national sporting venues. Others have found it much harder to win the affections of an entire nation. Stadium Australia, for example, which was built for the **Sydney Games** of 2000, is now better known as the Olympic Stadium but bears the official title of the ANZ Stadium, after a prolonged struggle over naming rights. It has been reconfigured, losing capacity in the process, to accommodate cricket and Australian Rules Football. It is unlikely ever to be regarded as a truly national stadium by citizens of other parts of the country.

St Louis Games (1904)

The third Olympiad of the modern era was the first Olympic Games held in North America, in St Louis, the USA's fourth most populous city at that time. Although initially Chicago had won the right to host the 1904 Summer Olympics, the organizers of the Louisiana Purchase Exposition in St Louis did not want to have another international event taking place at the same time, began to plan for their own sporting events and informed the Chicago organizing committee about their intentions. The expo organizers intended to eclipse the Olympic Games unless these were moved to St Louis. Eventually, in order to protect the **Olympic Movement**, Pierre de **Coubertin** gave in and awarded the Games to St Louis.

In the end, St Louis proved to be an unfortunate choice as a host city because the 1904 Olympics experienced precisely the same difficulty that had confronted Paris four years earlier, i.e. being overshadowed by a world fair. Indeed, the St Louis Olympics were reduced to a side-show of the exposition and were lost in the chaos of other cultural activities. Moreover, the time frame of the Games, with competitions being spread over four and a half months, again resulted in limited public interest. The number of attendees was further reduced by Europe-wide tension caused by the Russo–Japanese War (10 February 1904–5 September 1905), coupled with the logistical problems involved in getting to St Louis. Even de Coubertin failed to attend these Games, a significant statement in itself.

The 1904 Games hosted 651 athletes (645 men and 6 women), coming from 12 countries. Some 18 disciplines, comprising 16 sports, formed the Olympic programme. However, only 42 events (less than one-half) actually included athletes who were not from the USA.

These were the Olympics at which boxing, dumbbells, freestyle wrestling and decathlon made their debut appearances. One of the most remarkable athletes of the Games was the American gymnast, George Eyser, who won six medals even though his left leg was made of wood. Another interesting story concerns the marathon. Felix Carbajal, a Cuban postman, had to run in street clothes that he had cut around the legs to make them look like shorts. During the race he stopped at an orchard to have a snack on some apples, which turned out to be rotten. Those bad apples forced him to take an additional, unplanned break. Despite having to take a nap during the actual marathon, he still finished in fourth place.

In the end, the Olympics were turned into something of an experimental freak show, as the organizers decided to hold so-called *Anthropology Days* on 12 and 13 August. These involved various indigenous men from around the world, who were at the world's fair as parts of the exhibits, competing in a range of events so that anthropologists could see how they compared with the white man. Thus, the marathon included the first two black Africans to compete in the Olympics—two Tswana tribesmen, named Len Tau (real name, Len Taunyane) and Yamasani (real name, Jan Mashiani). However, they had not come to St Louis to compete in the Olympics, having been brought to the USA by the exposition organizers as part of the Boer War exhibit. The inclusion of this particular circus act and events of a similar nature was directly against the fundamental principles of the Olympic ethos and **human rights**.

Stockholm Games (1912)

Subsequently described as the 'Swedish masterpiece', the Games of the V Olympiad were hailed by Avery **Brundage** for the efficiency of their organization. Every continent was represented, thereby giving greater potency to the symbolism of the **Olympic Rings**. Automatic timing for track events was introduced for the first time, as was a public address system. In addition, the entire event took place in the space of one month rather than over a more extended period as had been customary. This was the last occasion on which 'private' entrants were permitted as opposed to representatives of national teams. It was the first time that an art competition was held as part of the Games, a practice that was discontinued after 1948.

In terms of controversy, the Swedish organizers did not allow boxing to be part of the Games. This led to a subsequent **International Olympic Committee (IOC)** ruling that hosts should not be allowed to control the programme of the Games in such a manner. Jim Thorpe of the USA, one of the great Olympians to have originated from **aboriginal people**, who won both the pentathlon and the decathlon events, was then disqualified for having contravened the amateur rules. This decision was not overturned until 1982.

The games took place in Stockholm's Olympic Stadium designed by architect Torben Grut. The impressive brick-built arena, surrounded by sculptures and close to the sports university, is still in regular use for a variety of sporting events and concerts. In 1956 it played host to the equestrian events of the **Melbourne Games**, quarantine rules having made it impossible for this element of the Games to go ahead in Australia.

Sydney Games (2000)

The Games of the XXVII Olympiad also became known as 'the Millennium Games' or 'the Games of the New Millennium'. They were further described by Juan Antonio **Samaranch** as 'the best Olympic Games ever' despite the fact that his wife had died during the course of the event. Many have disputed Samaranch's claim, not least those who have calculated the damaging social impact of the Games in terms of cost, democratic accountability, equity, the environment, employment, housing and legacy. The **International Olympic Committee (IOC)**, on the other hand, took comfort from two particular moments. First, the teams of the Democratic People's Republic of Korea (North Korea) and the Republic of Korea (South Korea) took part together in the opening ceremony, entering the stadium behind a unification flag. Second, Cathy **Freeman**, an Australian aboriginal athlete, won the 400 metres final. Freeman thus became the first person to light the Olympic flame and then go on to win a gold medal at the same Games. Her triumph was presented as symbolic not only of a more inclusive Australia but also, implicitly and as in the example set by the two Koreas, of the power of sport to unite people and to heal divisions.

Sydney won the right to host the Games in the face of competition from Istanbul, Berlin, Beijing and Manchester. Only one IOC member country was not represented, with Taliban opposition to sport ensuring that Afghanistan did not take part.

T

Terrorism

Just as peaceful political protesters are attracted to the Olympics and the publicity they engender (see **Protests**), so too are those committed to more violent modes of political expression. To date, the best-known example of this occurred at the **Munich Games** of 1972 when members of **Black September**, a militant Palestinian organization founded in 1970, murdered 11 Israeli athletes and officials and one German police officer. Holding nine of the Israeli team as hostages, the group demanded the release of over 200 political prisoners detained in Israel together with two German prisoners, Andreas Baader and Ulrike Meinhof, the founders of the terrorist Red Army Faction. More recently, on 27 July 1996, a bomb planted by Eric Rudolph exploded in the Centennial Olympic Park during the **Atlanta Games**, killing one person and injuring over 100 others. Unlike those of Black September, Rudolph's political objectives were incoherent, although right-wing in character. He had previously been involved in bomb attacks on a lesbian nightclub and on abortion clinics. Fears of a major terrorist threat to the 2008 **Beijing Games** were not fully realized. However, during the Games, terrorist bombs went off in the west of the country, the work of the Turkistan Islamic Party which demands independence for the Chinese province of Xinjiang. Previously, on 7 July 2005, four bombs had exploded on the London transport system, killing 52 commuters and four suicide bombers with probable links to Al-Qa'ida. The bombings took place the day after the **International Olympic Committee** had announced that London had won the right to host the Olympic Games in 2012. The timing may well have been coincidental; however, the bombings demonstrated that **security** would be a major issue for the London 2012 organizers. Indeed, speaking in 2006, London's Metropolitan Police Commissioner, Ian Blair, stated: 'There can be no doubt that the 2012 Games, if the current threat scenario remains the same, will be a huge target and we have to understand that and work on that basis'. With British troops still serving in Afghanistan, it can legitimately be suggested that the 'current threat scenario' does indeed remain.

Tokyo Games (1964)

Tokyo had originally been awarded the right to host the Summer Games of 1940. After Japan's invasion of China, however, the Games were reassigned to

Helsinki and then subsequently cancelled completely as a consequence of the outbreak of the Second World War. Having fended off the combined challenge of Detroit, Brussels and Vienna, Tokyo became the host city for the Games of the XVIII Olympiad which took place in 1964. In so doing, it became the first Asian city to host the Olympic Games, setting an example which was later to be emulated by Seoul in 1988 and Beijing in 2008. Hosting the Games was regarded by many as the culmination of Japan's post-war reconstruction efforts and an integral part of the rebuilding of the country's capital city. It was also taken as a symbol of Japan's acceptance by the international community and a response to the country's emergence as a major economic power. Judo was introduced to the Games for the first time at the request of the host nation and the Olympic flame was lit by Yoshinori Sakai, a student who had been born in Hiroshima on 6 August 1945, the day that the USA dropped an atomic bomb on the city. South Africa was banned from taking part and the Democratic People's Republic of Korea (North Korea) and Indonesia both withdrew. Malaysia competed for the first time after the political union of Malaya, British North Borneo, Sarawak and Singapore which had competed independently in 1960 and would do so again in the future. In athletics, Abebe Bikila, who in 1960 had become the first black African to win a gold medal, repeated his success in the marathon.

Torch relay

Although lighting the Olympic flame and the torch relay are modern Olympic rituals, fire had mythical connotations in ancient Greece as it was believed that Prometheus (a Titan) had stolen fire from the gods and given it to the people for whom it became the symbol of human reasoning, freedom and creativity. For his act, Prometheus was severely punished by being tied to a rock and his liver eaten daily by a vulture, only to be regenerated by night, due to his immortality, thus causing him eternal suffering. However, although fire, because its spiritual connotations, was present in many temples in ancient Olympia and was part of the Olympic sacrificial celebration, the torch relay, a modern addition, can be traced back only as far as the **Berlin Games** and particularly to Carl Deim, member of the **International Olympic Committee** and Secretary-General of the Organizing Committee of the 1936 Olympics. Deim believed that the torch relay would serve two fundamental purposes. It would showcase the strength of the German Aryan race and it would reinforce the claimed genetic link between ancient Greece and Germany. This idea swiftly captured the attention and approval of both Adolf Hitler and Joseph Goebbels, who wanted to use every opportunity to promote Nazism. The first torch relay, filmed by Leni Riefenstahl's documentary entitled *Olympia*, took place between Olympia and Berlin, the Olympic flame being carried by 3,331 runners (the number varies from source by source) through seven countries.

Months before the Games, the torch relay begins with the Olympic flame being ignited by the sun's rays, using parabolic mirrors, in Olympia in front of the ruins of the Temple of Hera. The flame is then placed in an urn and transported to the ancient stadium where it is given to the first runner. (Since 1964 the torch has been lit in the same way for the **Winter Games**, with one difference: the handing over of the flame to the first runner takes place near the monument to Pierre de **Coubertin**.) The flame is carried from the stadium to its final destination, the city hosting the Olympics. It is accepted practice that the flame is carried by a chain of runners. However, other means of transportation have also been used. For example, in 1976 the flame was transformed into a radio signal transmitted by satellite from Athens to Canada, where it was used to trigger a laser beam to re-light the flame. The torch relay ends on the day of the opening ceremony in the main stadium of the Games. The identity of the final carrier (a mystery runner who is usually a significant sports personality from the host country) is often kept secret until the last moment. This final bearer of the torch runs towards the cauldron and then uses the torch to ignite the flame in the stadium itself. After being lit, the flame continues to burn throughout the Olympic Games.

The torch relay preceding the **Beijing Games** became a major focus for international protests directed towards China's foreign and domestic policies (see **Protest**). In addition, the Taiwanese authorities refused to allow the torch to be carried through Taiwan because there was a feeling there that the Chinese were seeking to incorporate Taiwan into the domestic leg of the torch's journey.

V

Vikelas, Demetrios (1835–1908) (IOC President: 1894–96)

The first **International Olympic Committee (IOC)** President was born in Ermoupolis, on the island of Syros, on 15 February 1835. He grew up in Constantinople (now Istanbul) and later moved to London, where he became a successful businessman and married. As a consequence of his wife's mental illness, he moved to Paris and eventually gave up business to devote his life to the study of literature and history. During this time, he often travelled between France and Greece, these travels depending on his wife's health status. By the time he was approached to represent the Pan-Hellenic Gymnastic Club at the conference held at the Sorbonne in Paris, Vikelas had already earned a distinguished reputation in intellectual circles. He became the first IOC President and was instrumental in making the decision regarding the venue for the first Olympic Games. Even though Pierre de **Coubertin** would have preferred Paris to Athens as host city of the first modern Olympics, Vikelas's argument concerning Greece's historical links to the origins of the Olympics persuaded IOC members that Athens would be the most appropriate location. After the **Athens Games (1896)**, Vikelas resigned as IOC President and de Coubertin took over that position, subsequently holding it for over 29 years. Vikelas lived in Athens until his death in 1908.

W

Winter Games

Although ice-skating was proposed to be part of the Olympic programme from the outset of the Games and, indeed, figure skating was part of the **London Games (1908)** and the 1920 **Antwerp Games**, the first actual Winter Games were held in 1924 in Chamonix, France. After the Antwerp Games in 1920, the **International Olympic Committee (IOC)** came to the decision that the organizers of the **Paris Games (1924)** would also host a separate 'International Winter Sports Week'. This was to test how the public would receive the organization of such an event. The first Winter Games proved to be a great success, attracting more than 200 athletes from 16 nations, competing in 16 events. (It is important to note that the IOC only retrospectively, in 1926, recognized the Chamonix Winter Sport Week as the First Winter Olympic Games.)

The second Winter Games were held in St Moritz, Switzerland, in 1928. They attracted international attention but were adversely affected by warm weather. Consequently, the 10,000 metres speed skating event was abandoned and the 50 kilometres cross country event ended with the temperature being approximately 25°C, forcing many of the athletes to finish the race early.

The third Winter Olympics were awarded to Lake Placid in New York state, USA. However, owing to the Great Depression and the expense of the journey, fewer athletes participated in 1932 than in 1928. These Games were also affected by warm weather, leading to a two-day extension to the event. At this Olympics, American boxer, Eddie Eagan, won gold in the bobsled event, thereby becoming the only athlete to win gold medals in both the Summer and Winter Games.

In 1936 the twin towns of Garmisch and Partenkirchen in Bavaria joined forces to organize the fourth Winter Games. Alpine skiing made its Olympic debut in Germany, but skiing teachers were barred from entering the event as they were considered professionals. This IOC decision led to the first Winter Games-related **boycott**—by Swiss and Austrian skiers.

The outbreak of the Second World War significantly affected the Olympics and the 1940 and 1944 Games were cancelled (see **Cancelled Olympics**). Originally, the 1940 Winter Olympics were to have taken place in Sapporo, Japan, whilst the 1944 Winter Olympics were planned for Cortina d'Ampezzo in Italy. The **Olympic Movement** regrouped, albeit on a relatively small scale, in 1948 in St Moritz, Switzerland, with the fifth Winter Games. Although 28 countries competed at St Moritz, athletes from Germany and Japan were not invited due to problematic post-war diplomatic relations.

In 1952, the Winter Games were awarded to Oslo, Norway. The country is considered the birthplace of modern skiing and as a tribute to Sondre Nordheim, a skiing pioneer and the inventor of telemark skiing, the Olympic flame was lit in the fireplace of his home. Although only represented by West German athletes, Germany returned to the Olympic Games after 16 years.

In 1956 Cortina d'Ampezzo, Italy, organized the sixth Winter Olympics, the first to be televised. These Games witnessed the debut appearance of the USSR at the Winter Olympics, its athletes immediately showing their sporting prowess by winning more medals than any other nation.

When Squaw Valley, California, was awarded the Winter Olympics in the late 1950s, the place was little more than a 'ghost town'. After the IOC's decision to hold the 1960 Winter Games there, a rush of construction projects began, especially traffic infrastructure, hotels, restaurants, an ice arena, a speed skating track, ski lifts and a ski jumping hill. By the year of the opening ceremony, everything was in working order. Biathlon made its debut at these games. It is worthy of note that although biathlon first appeared in the Olympic programme in 1960, an event known as 'military patrol', derived from exercises carried out by Norwegian soldiers and involving cross country skiing and target shooting, had been part of previous Winter Games. This sport, briefly dropped after the Second World War due to post-war anti-military sentiments, was revived as the biathlon in 1955, and is now an integral part of the Winter Olympics.

Innsbruck in the Tyrol region of Austria staged the 1964 Winter Olympics. Despite careful organization, snow was the one thing that could not be planned with 100% certainty and this was indeed to hinder the Innsbruck Games. However, the organizers decided to defy nature's inefficiency in producing snow and ice and called upon the Austrian army to transport these 'products' to the sport venues. Luge (a small sled, capable of carrying one or two people) was first introduced in these Games, although the sport received bad publicity when a competitor was killed in a pre-Olympic training run.

Grenoble, France, won the right to host the 1968 Winter Olympics and witnessed some important historic milestones in Olympic history. That year, the IOC first permitted the German Democratic Republic (East Germany) and the Federal Republic of Germany (West Germany) to compete as separate nations at the Olympics, and drug- and gender-testing of competitors were made compulsory for the first time. Norway won the most medals at these Games, an achievement that had become the preserve of the USSR since it had first entered the Winter Olympics in 1956.

The 1972 Winter Games were held in Sapporo, Japan, the first time outside of North America or Europe. These Games were tainted by arguments revolving around the issue of professionalism. Three days before the Olympics, IOC President Avery **Brundage** threatened to bar a large number of top alpine skiers from competing because they did not comply with the strict IOC **amateur code**. Eventually, only Austrian star Karl Schranz, who presumably made the most money of all the skiers, was not allowed to compete. The amateur–professional tension, however, did not end at these Games and would haunt other Winter and Summer Olympics in the years that followed.

In 1980 the Winter Games returned to Lake Placid to face a range of political issues. At these Games, the People's Republic of China made its debut, leading the Republic of China (Taiwan) to **boycott**. The IOC forced the Republic of China (Taiwan) to compete under the name of Chinese Taipei, which the Taiwanese refused, and thus became the only nation to boycott the Olympic Winter Games. The threat of the American boycott of the 1980 Summer Olympics also affected these Olympics and the **Cold War**-generated tension filtered into some of the events. A case in point was the Olympic ice-hockey tournament, more specifically the match between the Soviet professional and American amateur sides. The Soviets were believed to be unbeatable, but the Americans staged an outstanding performance and won the match. (This match was later dubbed the *Miracle on Ice*, also the title of a Walt Disney movie about the game that was directed by Gavin O'Connor and released in 2004.) The American team went on to win the Olympic title by beating the Finnish team.

In 1984 the Winter Games were staged in Sarajevo of the Socialist Federal Republic of Yugoslavia (Sarajevo is now the capital city of Bosnia and Herzegovina, a country that was established after the break-up of Yugoslavia in the 1990s). It was the first time that the Winter Olympics had taken place in a socialist country and is still the only Olympics to have been held in a predominantly Muslim city. The Republic of China finally ended its boycott over the controversy regarding the IOC's recognition of the People's Republic of China and competed as Chinese Taipei for the first time.

Calgary in Canada was awarded the right to host the 1988 Winter Olympics. An interesting feature of these Games was the appearance of Jamaica's first ever bobsleigh team which received considerable attention. (In 1994 a film entitled *Cool Runnings*, directed by Jon Turteltaub, captured the heroic efforts of these sporting underdogs in a touchingly sentimental fashion.)

In 1986 the IOC decided to reschedule the Summer and Winter Games by alternating between them every two years. This meant that each would still be held in four-year cycles, but two years apart from one another. The 1992 Albertville Games were the last to be held in the same year as the Summer Games. Since these Winter Games took place only a few years after the collapse of communist regimes in many Central and Eastern European countries, political change was reflected in the teams that took part. For example, Germany competed as a united nation and Croatia and Slovenia, nations of the former Yugoslavia, made their debuts as independent states. Most of the former Soviet republics still competed as a single team, under the name of Unified Team, but the Baltic states participated independently for the first time since before the Second World War.

Lillehammer, Norway, hosted the 1994 Winter Olympics. After the break-up of Czechoslovakia in 1993, the Czech Republic and Slovakia made their Olympic debuts as independent countries, as did several other states that had split from the former USSR. These were arguably the first Olympics, Summer or Winter, at which **environmentalism** was treated as a serious issue for international sport.

In 1998 the Japanese city of Nagano hosted the Winter Olympics. Indicating the growing global popularity of the Winter Games, the number of participants exceeded 2,000. Women's events were extended, with the inclusion of ice-hockey, and the men's ice-hockey event finally became open to players with both professional and amateur status, leading to a huge influx of National Hockey League (NHL) players.

In 2002 Salt Lake City, Utah, staged probably the most infamous Winter Olympics to date. Prior to the opening of the Games, it was found that the organizers had bribed several IOC members in order to secure the right to host the 2002 Games (see **Bidding**). This issue has had a long-term and fundamental effect on the host city selection process and on the IOC, members of which had to resign after the emergence of the Salt Lake City scandal. Another difficulty with regard to organizing the event was that the Salt Lake City Games were the first Olympics since 11 September 2001, which meant a need for tight **security** measures amidst a potentially high threat of terrorist attack.

At the Games themselves, controversy occurred in connection with the pairs figure skating contest. Canadians Jamie Salé and David Pelletier were awarded silver for their performance. Although the athletes accepted their placing with forbearance, the crowd and the media protested. The next day it came to light that the French jury member had been pressurized into favouring the Russian pair and, consequently, the IOC and the International Skating Union decided to discard the vote, suspend several judges, and award both pairs the gold medal. The scandal resulted in a fundamental change to the scoring system used for figure skating.

Turin, Italy, earned the right to organize the 2006 Winter Games. Although somewhat different in nature, these Olympics were not without problems. These included bankruptcy threats, doping issues, poor attendance and an unfinished transportation system. Security was another issue, which the organiz ers further increased following Denmark's *Jyllands-Posten* Muhammad cartoons controversy. Fortunately, the Games concluded without a major breach of security. The next Winter Olympic Games will be held in Vancouver, Canada in 2010, followed by Sochi in Russia in 2014.

Z

Zátopek, Emil (1922–2000)

Born on 19 September 1922 in Koprivnice, Czechoslovakia, Zátopek won gold and silver medals in the 10,000 metres and 5,000 metres, respectively, at the **London Games** in 1948. In 1952 in Helsinki, he surpassed this achievement by winning gold medals in both events and in his first ever marathon, setting new Olympic records in each. Nicknamed 'the Czech locomotive', he is credited with having revolutionized training techniques (legend has it that he used to run in army boots whilst carrying his wife on his back). He served in the Czech army and was promoted to the rank of colonel. However, in 1968, as a member of the liberal wing of the Czech Communist Party, Zátopek signed First Secretary Alexander Dubček's manifesto in support of liberal reforms and subsequently opposed the Soviet invasion of his country. He was expelled from the Communist Party, dismissed from the army and forced to work on a geological survey in a uranium mine. With the collapse of communism in Czechoslovakia, Zátopek's army rank was restored and his athletic achievements were recognized in 1998 with the award of the Order of the White Lion from President Vaclav Havel. He died, following a stroke and several years of ill health, on 22 November 2000.